T0262356

Biomaterials: Scientific Aspects

Biomaterials: Scientific Aspects

Edited by **Ralph Seguin**

New York

Published by NY Research Press,
23 West, 55th Street, Suite 816,
New York, NY 10019, USA
www.nyresearchpress.com

Biomaterials: Scientific Aspects
Edited by Ralph Seguin

© 2015 NY Research Press

International Standard Book Number: 978-1-63238-064-7 (Hardback)

Contents

Preface

This book was inspired by the evolution of our times; to answer the curiosity of inquisitive minds. Many developments have occurred across the globe in the recent past which has transformed the progress in the field.

This book consists of reviews and original researches conducted by experts and scientists working in the field of biomaterials, its development and applications. It offers readers with the potentials of distinct synthetic and engineered biomaterials. This book gives a comprehensive summary of the applications of various biomaterials, along with the techniques required for designing, developing and classifying these biomaterials without any intervention by any industrial source. The book deals with latest researches on new and known materials and gives special attention to their physical, chemical and mechanical characteristics, and also discusses biocompatibility and histopathological studies. It also elucidates the various techniques used to produce biomaterials with the required physical and biological features for medical and clinical applications.

This book was developed from a mere concept to drafts to chapters and finally compiled together as a complete text to benefit the readers across all nations. To ensure the quality of the content we instilled two significant steps in our procedure. The first was to appoint an editorial team that would verify the data and statistics provided in the book and also select the most appropriate and valuable contributions from the plentiful contributions we received from authors worldwide. The next step was to appoint an expert of the topic as the Editor-in-Chief, who would head the project and finally make the necessary amendments and modifications to make the text reader-friendly. I was then commissioned to examine all the material to present the topics in the most comprehensible and productive format.

I would like to take this opportunity to thank all the contributing authors who were supportive enough to contribute their time and knowledge to this project. I also wish to convey my regards to my family who have been extremely supportive during the entire project.

Editor

New Materials for Biomedical Applications: Chemical and Engineering Interventions

1

Galectins: Structures, Binding Properties and Function in Cell Adhesion

Christiane E. Römer and Lothar Elling
Helmholtz-Institute for Biomedical Engineering, RWTH Aachen University
Germany

1. Introduction

Galectins are nearly ubiquitous distributed β-galactoside binding proteins which share a common amino acid sequence, the carbohydrate recognition domain (CRD) (Barondes et al., 1994a; Cooper, 2002; Elola et al., 2007; Hirabayashi & Kasai, 1993; Hughes, 2001; Klyosov et al., 2008; Leffler et al., 2004). They are evident in vertebrates, invertebrates and also protists, implying fundamental functions of these lectins (Hirabayashi & Kasai, 1993). Some galectins are distributed in a variety of different tissues, others are more specifically expressed (Cooper, 2002).

Galectins are known to perform high diversity of functions inside the cells and in the extracellular space. They are regulators of cell cycle, inflammation, immune responses, cancer progression, cell adhesion, cell signalling events and so on. The different functions are performed either by protein-protein or by protein-glycan interactions (Almkvist & Karlsson, 2004; Danguy et al., 2002; Elola et al., 2007; Hernandez & Baum, 2002; Hughes, 2001; Ilarregui et al., 2005; Liu et al., 2002; Liu & Rabinovich, 2005; Rabinovich et al., 2002b; Rabinovich & Toscano, 2009; Vasta, 2009).

Different excellent reviews focus on the wide-spread functions of galectins such as tumor progression, cell signalling or inflammation (Garner & Baum, 2008; Hernandez & Baum, 2002; Liu et al., 2002; Liu & Rabinovich, 2005; Nangia-Makker et al., 2008; Rabinovich et al., 2002a; Rabinovich & Toscano, 2009; van den Brule et al., 2004; Vasta, 2009). Review articles discussing functions of galectins in cell adhesion events and their role as matricellular proteins for the crosslinking of extracellular matrix components have also been published (Elola et al., 2007; Hughes, 2001). The function of galectins in the assembly of the extracellular matrix as well as in cell adhesion and cell signalling processes shows their potential as mediators for cell adhesion and proliferation on biomaterial surfaces. Galectins are interesting candidates for the functionalisation of biomaterial surfaces as they can promote the primary binding event of cells to foreign materials and influence specific signalling processes. In this article we want to analyse the potential use of galectins (explained by the examples of galectin-1, -3 and -8) in biomaterial research and application.

2. Families and structures of galectins

Galectins are defined by their β-galactoside binding ability and their common sequence of about 130 conserved amino acids. This sequence homology results in a similar overall three-dimensional structure of the carbohydrate recognition domain (CRD) (Barondes et al., 1994a; Barondes et al., 1994b). Several human galectin CRDs have been characterised by crystallography, including those of human galectin-1, galectin-3 and the N-terminal domain of galectin-8 (Ideo et al., 2011; Kishishita et al., 2008; Lobsanov et al., 1993; Lopez-Lucendo et al., 2004; Seetharaman et al., 1998). The C-terminal domain of galectin-8 has been investigated by NMR (Tomizawa et al., 2008). All of them show a globular fold consisting of two anti-parallel β-sheets with five to six strands respectively (Ideo et al., 2011; Kishishita et al., 2008; Lobsanov et al., 1993; Lopez-Lucendo et al., 2004; Seetharaman et al., 1998; Tomizawa et al., 2008). The CRDs analysed so far consist of three consecutive exons, with most of the conserved amino acids encoded on the middle one (Cooper & Barondes, 1999; Houzelstein et al., 2004).

2.1 Galectin families
Regarding their overall structure galectins are clustered in three families: a) prototype galectins consisting of one CRD, b) chimera-type galectins with one CRD and a non-lectin domain whose only member known so far is galectin-3, and c) tandem-repeat galectins which have two different CRDs linked by a short peptide (see Fig. 1) (Hirabayashi & Kasai, 1993; Leffler et al., 2004).
In this review we want to focus on galectin-1, -3 and -8 as representatives of the three galectin families.

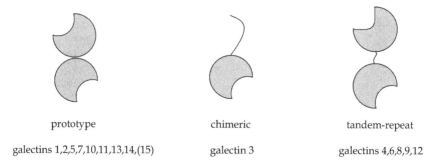

| prototype | chimeric | tandem-repeat |

galectins 1,2,5,7,10,11,13,14,(15)   galectin 3   galectins 4,6,8,9,12

Fig. 1. Galectin families regarding their overall structure (modified from Barondes et al. 1994) (Al-Ansari et al., 2009).

Galectins are either divalent regarding their intrinsic protein structure (tandem-repeat galectins such as galectin-8) or form homotypic di- to oligomers through site-specific interactions (prototype and chimera galectins such as galectin-1 and -3). Different galectin-8 isoforms represent either prototypic or tandem-repeat type galectins depending on splice variants. Some galectin-8 splice variants consist only of the N-terminal CRD and different elongations without a second CRD. Those variants can rather be grouped to the prototypic galectins (Bidon et al., 2001; Al-Ansari et al., 2009 Zick et al., 2004). Prototype galectins such as galectin-1 form homodimers through hydrophobic interactions at the N-terminal amino acid residues (Cho & Cummings, 1997; Lobsanov et al., 1993). Dimerisation occurs as equilibrium reaction depending on protein concentration but independent of available soluble ligands (Cho & Cummings, 1995). In contrast the chimeric galectin-3 forms oligomers (most likely pentamers) via its N-terminal collagen-like extension after ligand-binding (Ahmad et al., 2004a; Birdsall et al., 2001; Nieminen et al., 2008).

2.2 Carbohydrate recognition domains

The carbohydrate recognition domain is highly conserved throughout different galectins and organisms.

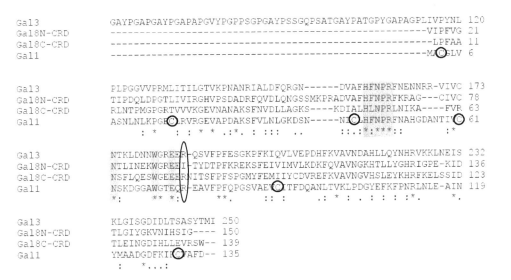

Fig. 2. Sequence alignment of human galectin-1, -3 and the single carbohydrate recognition domains of galectin-8 (residues 1-150 and 221-359 of isoform a). Used sequences are galectin-1 (NP_002296), galectin-3 (P17931) and galectin-8 isoform a (NP_963839) as published on http://www.ncbi.nlm.nih.gov/protein. Completely conserved amino acids are marked with an asterisk, conserved substitutions are marked with a colon and semi-conserved substitutions with a simple dot. Important amino acids mentioned in the following text are additionally highlighted: Conserved amino acids of the binding pocket are highlighted in grey; residues with importance for the binding are labelled in an ellipse; galectin-1 cysteine residues are marked with circles. The alignment has been performed using ClustalW2 at http://www.ebi.ac.uk using the default settings (Chenna et al., 2003; Larkin et al., 2007).

The conserved amino acids are directly involved in carbohydrate binding either by the formation of hydrogen bonds or van der Waals interactions with the sugar moiety. Most of the conserved amino acids form hydrogen bonds with the bound sugar unit. An important sequence motif in this context is His(158)-Asn(160)-Arg(162) (numbering according to human galectin-3, see Fig. 2). Those three amino acids have been found to form hydrogen bonds with the bound galactose residue for example in galectin-1 (Lobsanov et al., 1993), galectin-3 (Diehl et al., 2010; Seetharaman et al., 1998), and galectin-8 N-CRD (Ideo et al., 2011; Kishishita et al., 2008) (see Fig. 3). The sequence motif can also be found in galectin-8

C-CRD but as no x-ray crystallography is available for this CRD the hydrogen bridges have not been verified yet. Additional residues are involved in the conserved binding process either by hydrogen bonding (Glu184, Asn174, numbering according to human galectin-3, see Fig. 3) or van-der-Waals interaction (Trp181, numbering according to human galectin-3) (Di Lella et al., 2009; Diehl et al., 2010; Lobsanov et al., 1993; Seetharaman et al., 1998).

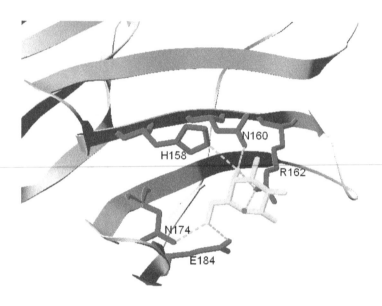

Fig. 3. Human galectin-3 with bound galactose unit of LacNAc **PDB 1KJL**. H-bondings are shown as dotted lines for residues H158 (C4-OH), N160 (C4-OH), R162 (C4-OH and intramolecular O-atom), N174 (C6-OH), E184 (C6-OH) (Seetharaman et al., 1998; Sörme et al., 2005). Picture made with SwissProt pdb viewer (Guex & Peitsch, 1997).

The importance of the mentioned H-bonding amino acids has been proven by site-directed mutagenesis performed with human galectin-1. In those experiments the change of single amino acids involved in H-bonding eliminates the binding to lactose-sepharose and/or asialofetuin (Hirabayashi & Kasai, 1991; Hirabayashi & Kasai, 1994). Although binding is not completely abolished, significant influence of the conserved Trp residue for sugar binding was also proven in bovine and human galectin-1 (Abbott & Feizi, 1991; Hirabayashi & Kasai, 1991).

Arg186 is not completely conserved throughout the different galectins (see Fig. 2, elipse). The N-terminal domain of galectin-8 for example presents an Ile at the corresponding position resulting in a differing fine specificity for glycans. Due to this mutation galectin-8 N-CRD favours lactose structures over LacNAc type II structures in the binding site. Thereby different biological functions of galectin-8 in contrast to other galectins such as galectin-3 are regulated (Salomonsson et al., 2010). Specific binding preferences resulting from the differences in amino acid sequence will be discussed in chapter 4 in more detail.

2.3 Other specific features of galectins
2.3.1 Secretion

Different tissues are known to produce galectins and most of them secrete parts of the cytosolic galectin pool. The amount of secreted galectin depends on cell type, differentiation status and can be regulated by external triggers (Cooper, 2002; Hughes, 1999). Examples of galectin producing cells with relevance for regenerative medicine are beside others neurons, epithelial cells of several tissues and liver cells, which produce either several different galectins or a specific subset of galectins (Dumic et al., 2006; Hughes, 1999).

Galectins act intra- and extracellularly. As known so far they are secreted via a non-classical mechanism which is not fully understood yet. They lack classical signalling sequences for specific localisation but can be found in the outer cellular space as well as inside the cells even located in the nucleus (Hughes, 1999). Although the complex regulation of secretion remains still elusive some explanations have been found: Galectin-1 secretion depends on the binding to a counter-receptor molecule and does not involve plasma membrane blebbing (Seelenmeyer et al., 2005; Seelenmeyer et al., 2008). Galectin-3 secretion seems also to be regulated by binding to other proteins such as chaperons and subsequent vesicular secretion (Hughes, 1999; Mehul & Hughes, 1997). The N-terminal-domain of galectin-3 is important for subcellular translocation and secretion of the protein (Gong et al., 1999).

2.3.2 Galectin-1: Importance of reducing conditions

The lectin activity of galectin-1 depends on reduced cysteine residues. Oxidised galectin-1 has no lectin activity but functions in the regeneration of nerve axons (Horie et al., 2004). Galectin-1 has six cysteine residues which are accessible to the solvent (see Fig. 2). The removal of the most accessible cysteine (Cys2) (Lopez-Lucendo et al., 2004) - or better all cysteine residues - enhances protein stability under both reducing and non-reducing conditions significantly (Cho & Cummings, 1995; Nishi et al., 2008), while none of them is necessary for lactose binding as shown by site directed mutagenesis and x-ray crystallography (Hirabayashi & Kasai, 1991; Lopez-Lucendo et al., 2004).

2.3.3 Galectin-3: The only known chimera type galectin

Galectin-3 has some specific properties due to its unique structure. Galectin-3 consists of three parts: 1) a N-terminal 12 amino acid leader sequence containing two phosphorylation sites, 2) a proline and glycine rich collagen like domain necessary for oligomerisation and 3) the carbohydrate recognition domain (Ahmad et al., 2004a; Dumic et al., 2006; Kubler et al., 2008; Mehul & Hughes, 1997; Nieminen et al., 2008). The first few amino acids forming the leader peptide are important for the subcellular localisation and secretion of the protein (Gong et al., 1999). Moreover phosphorylation of Ser6 seems to regulate affinity for different ligands and thereby cellular activity of galectin-3 (Dumic et al., 2006; Mazurek et al., 2000; Szabo et al., 2009; Yoshii et al., 2002). Galectin-3 can be cleaved by different proteases such as metalloproteinases-2 and –9 (gelatinases A and B respectively), metalloproteinase-13 (collagenase-3) and with low activity metalloproteinase-1 (collagenase-1) separating the full-length CRD from the N-terminal extension (Guévremont et al., 2004; Ochieng et al., 1994). The main cleavage position is located between Ala62 and Tyr63 while other cleaving sites are only recognised by some specific proteases to lesser extend (Dumic et al., 2006;

Guévremont et al., 2004; Ochieng et al., 1994). The single CRD is mainly described to have an increased affinity for different carbohydrates such as N-acetyllactosamine, the glycoprotein asialofetuin or glycans presented on endothelial cells but to have less biological activity as it looses the ability to form oligomers. This reveals the possible regulatory function of galectin-3 cleavage (Dam et al., 2005; Dumic et al., 2006; Ochieng et al., 1998a; Shekhar et al., 2004). In terms of this regulation it is suggested that the single galectin-3-CRD binds with high affinity to glycans on cell surfaces thereby blocking these interaction partners for full-length galectin-3 binding. After this blockage the full-length protein cannot perform its physiological functions anymore. In this way galectin-3 cleavage could act as down-regulation of galectin-3 function (John et al., 2003; Shekhar et al., 2004).

2.3.4 Galectin-8: Several isoforms of a tandem-repeat galectin
The specific properties of galectin-8 are also implied in its structure and the different isoforms arising from it. At least 6 different isoforms are described so far of which some only consist of the N-terminal CRD with an extension and others consist of both CRDs linked by different hinge domains (Bidon et al., 2001; Delgado et al., 2011; Zick et al., 2004). The two galectin-8 CRDs show approximately 35% sequence similarity but reveal different fine specificity for glycan structures. Therefore galectin-8 can act as "hetero-bifunctional crosslinking agent" (Zick et al., 2004). The length and structure of the linker domain has direct influence on the biological function (Levy et al., 2006). Moreover the linker domain regulates susceptibility to protease cleavage. It was for example shown that a long linker can be cleaved by thrombin while shorter linker variants are not substrate for this protease (Nishi et al., 2006).

3. Glycan binding assays for galectins

As galectins play a fundamental role in cell adhesion, cell signalling, inflammation, tumor progression etc. there is an enormous interest in the evaluation of galectin-glycan interactions regulating those functions.

3.1 Comparison of different common assays
Various assay set-ups have been designed to analyse the binding behaviour of different galectins to specific glycan structures. Binding assays can be subdivided regarding the presentation of the different binding partners: 1) the glycan structure is immobilised, 2) the galectin is immobilised and 3) both binding partners are soluble.
The chosen assay format influences the data generated as each assay set-up has its own advantages and disadvantages (Rapoport et al., 2008):
Assays in which one of the binding partners is immobilised raise the problem that the amount of this ligand is not completely known. Moreover it is possible that side interactions with the surface occur or that the conformation and flexibility of the bound partner differ slightly from its soluble parameters. The natural oligomerisation of galectins is blocked after immobilisation. Beside this the presentation of the immobilised binding partner is multivalent which influences the binding (Sörme et al., 2004). This can be useful for the glycan structures, as they are multivalently presented in nature as well, but not for galectins. Examples for studies with immobilised glycans or glycoproteins are glycan arrays, ELISA

assays and surface plasmon resonance (Appukuttan, 2002; Blixt et al., 2004; Bohorov et al., 2006; Ideo et al., 2003; Munoz et al., 2010; Song et al., 2009b; Stowell et al., 2008a). For glycan arrays it can be important which linker is used to bind the glycan epitopes to the surface (length, chemical structure). Moreover it is possible to use chemically and/or enzymatically produced ligands as well as glycans from natural compounds like glycopeptides and glycolipids (Blixt et al., 2004; Bohorov et al., 2006). The latter allows the analysis of complex and even unknown glycan structures of different cells (Song et al., 2009a; Song et al., 2009b; Song et al., 2010). Immobilised galectins are for example used in frontal affinity approaches and ELISA assays (Hirabayashi et al., 2002; Sörme et al., 2002).

Variations of binding assays with immobilised partners are assays in which the surface binding is inhibited by a soluble ligand. Such inhibition studies of surface interactions allow a direct read-out of IC50 values and thereby the direct comparison of relative affinities (Sörme et al., 2002). For the calculation of affinity constants assumptions have to be made to simplify calculations which might not be correct for each single interaction measurement. Additional the problems mentioned before still persist (Sörme et al., 2004).

Most assays with one immobilised component as well as some direct interaction assays are based on the read-out of a fluorescence signal or other labels. Therefore either the galectin or the glycan structures have to be labelled. This leads to some additional problems: If the glycan is chemically labelled the linker or label itself can alter the binding affinity with specific effects for different galectins (Sörme et al., 2004). Therefore the affinity constants measured do not exactly fit to the unmodified glycan structures. Moreover the labelling of glycans is time-consuming. The labelling of galectins can also alter the binding specificities. It is in most cases done by random chemical modification of specific functional groups such as amino or thiol functionalities (Carlsson et al., 2007; Patnaik et al., 2006; Rapoport et al., 2008; Salomonsson et al., 2010; Song et al., 2009b; Stowell et al., 2008a; Stowell et al., 2008b). Although this labelling is assumed not to influence binding specificity or inactive galectins are removed after the labelling reaction, binding and oligomerisation still might be slightly affected. Moreover lot-specific aberrations between different labelling reactions occur. Labelled galectins are for example used for glycan arrays and ELISA-type set-ups (Blixt et al., 2004; Carlsson et al., 2007; Rapoport et al., 2008; Salomonsson et al., 2010; Song et al., 2009b; Stowell et al., 2008a) while fluorescence labelled glycans are used in frontal affinity chromatography or fluorescence polarisation (Carlsson et al., 2007; Hirabayashi et al., 2002; Salomonsson et al., 2010; Sörme et al., 2004).

Assays using both binding partners in its soluble form overcome most of the mentioned problems. But although those assays have different advantages the results cannot directly be compared with the natural set-up in which the glycans are immobilised on glycoproteins or glycolipids and thereby multivalently presented. Fluorescence polarisation is one of these methods measuring direct interactions of ligands in solution, but facing negative side effects of glycan labelling. Similarly, titration calorimetry also measures the interaction of two soluble binding partners. As for titration calorimetry no labelling reaction has to be performed this assay set-up might be considered as the one with fewest problems. But needed galectin concentrations for this test are usually (but not always) in high micromolar ranges and therefore above the physiological range. In this concentration range galectins tend to oligomerise, aggregate or precipitate (Ahmad et al., 2004b; Bachhawat-Sikder et al., 2001; Cho & Cummings, 1995; Dam et al., 2005; Sörme et al., 2004). Moreover titration calorimetry experiments are suitable for comparative studies of different glycans but do not lead to accurate calculation of affinity constants (Ahmad et al., 2004b). Another way to determine the direct interaction of soluble galectins and glycans is the use of hemagglutination assays, but those are limited to multivalent glycans or the inhibition of interactions between galectins and multivalent glycans or erythrocytes (Ahmad et al., 2004a; Ahmad et al., 2004b; Appukuttan, 2002; Giguere et al., 2008).

3.2 Determined fine specificity of galectin-1, -3 and the two galectin-8 CRDs
Although the general fold of all galectins is highly conserved, single galectins are characterised by specific binding interactions with single carbohydrate ligands. Differences in fine specificity have been analysed using different binding assays (as mentioned above). Moreover extensive theoretical evaluation of the putative interactions between single amino acids and functional groups of the bound glycan has been done by modelling and calculation. Some specific ligands with high affinity for the single galectin CRDs are mentioned in Table 1.

The recognition of galactose is common for all galectins but the interaction with the monosaccharide alone is very weak (Carlsson et al., 2007; Knibbs et al., 1993; Salameh et al., 2010). Disaccharides containing galactose β-glycosidic bound to GlcNAc, Glc or GalNAc are bound with significantly increased affinities. Different galectins thereby show high affinity to specific disaccharides. Galectin-3, galectin-1 and the C-terminal CRD of galectin-8 bind preferentially LacNAc units of type I and type II while the N-terminal CRD of galectin-8 shows highest affinity for lactose (Carlsson et al., 2007; Ideo et al., 2011; Salomonsson et al., 2010).

Extensions of the bound galactose moiety effect glycan binding in dependence on the galectin. Galectin-3 tolerates due to its enlarged binding pocket extensions at the galactose 3`-OH-group for example repetitive LacNAc (type II) –structures (poly-LacNAc), showing even higher affinities for repetitive LacNAc structures compared to single LacNAc units (Hirabayashi et al., 2002; Rapoport et al., 2008; Salomonsson et al., 2010). In contrast galectin-1 recognises single LacNAc units presented at the non-reducing terminus of glycans not showing preference for extended poly-LacNAc glycans (Leppänen et al., 2005). Most authors agree that galectin-1 is not able to bind internal galactose moieties in poly-LacNAc-glycans (GlcNAc-β3-Gal-β4-GlcNAc) (Leppänen et al., 2005; Stowell et al., 2004; Stowell et al., 2008a) but depending on the assay set-up some publications report affinity to this sugar unit (Di Virgilio et al., 1999; Zhou & Cummings, 1993). These different results prove the importance of evaluation of the test set-up and critical examination of the measured binding data.

Other extensions at the 3`-OH-group of galactose such as sulphate or neuraminic acid increase the affinity of galectin-3, galectin-1 and especially galectin-8 N-CRD to the core disaccharide (Carlsson et al., 2007; Sörme et al., 2002; Stowell et al., 2008a). In contrast the C-terminal galectin-8 domain fails to bind 3`-sulfated or 3`-sialylated galactose (Ideo et al., 2003).

Modification at the 6`-OH-group for example with neuraminic acid reduces binding of all four discussed galectin CRDs (Ideo et al., 2003; Stowell et al., 2008a). Therefore α6-sialylation is discussed as regulatory modification for galectin-mediated functions (Zhuo & Bellis, 2011). Galectin-3 and galectin-8 C-CRD show high affinity for blood-group antigens (Hirabayashi et al., 2002; Yamamoto et al., 2008)

Galectin-1	Galectin-3	Galectin-8 N-CRD	Galectin-8 C-CRD
Complex N-glycans Increasing affinity with increasing number of antennas	N-glycans, preferred poly-LacNAc	N-glycans and glycosphingolipids (e.g. GM3 and GD1a)	N-glycans, preferred poly-LacNAc
Non-reducing terminal LacNAc type I or II (grey in upper scheme) β3/4	Non-reducing terminal and internal LacNAc, high affinity for repetitive LacNAc-units (grey in upper scheme) β4	Preference for lactose (but also binding to LacNAc-units) β4	LacNAc type I or II high affinity for repetitive LacNAc-units (grey in upper scheme) β3/4
3-O-sulfation SO$_4$ β3/4	3 O-sulfation SO$_4$ β4	3 O-sulfation and α3-sialylation SO$_4$ β4 α3 β4	
	Blood group A and B antigens α2 α3 β4 α α3 β4		Blood group A and B antigens α2 α3 β4 α2 α3 β4

Table 1. Preferred ligands of the single carbohydrate recognition domains of galectin-1, -3 and -8 following Rapoport et al. 2002. Symbols according to the consortium of functional glycomics (Brewer, 2004; Carlsson et al., 2007; Dell, 2002; Hirabayashi et al., 2002; Ideo et al., 2003; Ideo et al., 2011; Leppänen et al., 2005; Patnaik et al., 2006; Rabinovich & Toscano, 2009; Rapoport et al., 2008; Salomonsson et al., 2010; Stowell et al., 2004; Stowell et al., 2008a; Yamamoto et al., 2008)

4. Glycoproteins as binding partners of galectins

Galectins can act intracellular or in the extracellular space, where they have different functions regulated by protein-protein or protein-glycan interactions. In the extracellular space they interact with different glycoproteins influencing cell adhesion, signalling and proliferation events. Thereby they interact with ECM-glycoproteins forming the extracellular matrix and with glycosylated transmembrane or membrane associated proteins on the cell surface (table 2). Following we present some selected binding partners of the three different galectins discussed so far.

4.1 ECM glycoproteins as binding partners of galectins

Different extracellular matrix proteins contribute to structural and functional aspects of the extracellular space. Galectin-1 interacts strongly with different extracellular matrix proteins. It has affinity to several glycoproteins as with increasing affinity osteopontin, vitronectin, thrombospondin, cellular fibronectin and laminin (Moiseeva et al., 2003). Most of these interactions depend on the carbohydrate recognition domain and can be inhibited with soluble glycan ligands (Cooper, 1997; Moiseeva et al., 2003; Ozeki et al., 1995; Zhou & Cummings, 1993). Galectin-3 also shows high affinity for some ECM-glycoproteins (Dumic et al., 2006; Kuwabara & Liu, 1996; Massa et al., 1993; Matarrese et al., 2000; Ochieng et al., 1998b; Sato & Hughes, 1992). The best binding candidates fibronectin and laminin are heavily glycosylated (5-7% and at least 12-15% respectively), carrying mainly N-glycans (Paul & Hynes, 1984; Tanzer et al., 1993). N-glycans are among the main binding partners of galectin-1, -3 and –8 (Patnaik et al., 2006) (although galectin-8 also shows high affinity to some glycosphingolipids (Ideo et al., 2003; Yamamoto et al., 2008)). One third of laminin N-glycans is composed of repetitive "N-acetyllactosamine" units (shown for mouse EHS-laminin) which are preferentially recognised by galectin-3 but also by galectin-1 and to less extent galectin-8 (Arumugham et al., 1986; Hirabayashi et al., 2002; Knibbs et al., 1989; Sato & Hughes, 1992; Zhou & Cummings, 1993). The other ECM-glycoproteins carry also N-glycans but are less glycosylated (Bunkenborg et al., 2004; Chen et al., 2009; Liu et al., 2005). For example osteopontin from human bone shows only two N-glycans with binding sites which are partially blocked by α6-bound sialic acid (Ideo et al., 2003; Masuda et al., 2000; Stowell et al., 2008a). Another extracellular matrix protein interacting with galectin-1 and -3 is the Mac-2 binding protein or 90K antigen which influences adhesion processes (Sasaki et al., 1998; Tinari et al., 2001).

The different ECM-proteins which are bound by galectins can interact with other ECM-glycoproteins and/or integrins (Adams, 2001; Janik et al., 2010; Kariya et al., 2008; Singh et al., 2010). These interactions can lead to regulatory effects, lattice formation and signalling cascades.

4.2 Cell-surface glycoproteins as binding partners of galectins

Beside these soluble ECM components also some membrane-bound proteins are recognised by galectins. One of these is the lysosome associated membrane glycoprotein 1 (LAMP-1) which is known to carry several N-glycans partly presenting poly-lactosamine glycans recognised by galectin-3 and -1 (Chen et al., 2009; Do et al., 1990; Dong & Hughes, 1997). LAMP-1 is also known as CD107a. Several other membrane proteins associated in the cluster of differentiation such as CD3, 4, 7, 8, 43 and 45 which are presented on T-cells are also recognised by galectin-1, showing the function of galectin-1 in immune response and inflammation (Liu, 2005; Nishi et al., 2008; Pace et al., 1999; Rabinovich et al., 2002a; Rabinovich et al., 2002b).

Binding partner	Gal	Cell type	Process	Reference
Integrin α7β1	1	Skeletal muscle cells; Myoblasts	Influences integrin-laminin interaction	(Gu et al., 1994)
Integrin α1β1	1	Vascular smooth muscle cells	Influences adhesion and migration	(Moiseeva et al., 1999)
Integrin α3β1	8	e.g. endothelial cells	Influences cell adhesion and survival by modulating integrin-ECM interaction	(Hadari et al., 2000)
Cell recognition molecule L1; Myelin associated glycoprotein (MAG) Neural cell adhesion molecule (NCAM)	3	Neural tissue	Likely influences cell adhesion and signalling processes	(Probstmeier et al., 1995)
NG2 proteoglycan	3	Microvascular pericytes	Influences endothelial cell motility and morphogenesis	(Wen et al., 2006)

Table 2. Examples of cell-surface-glycoproteins interacting with galectins
This does not constitute a comprehensive list of cell-bound galectin-binding-glycoproteins, but just intends to show some examples which might be interesting for tissue engineering. Immune and tumor cells are not included in the list.

Similarly galectin-3 binds to CD98 on macrophages, CD66 on neutrophils and the T-cell receptor also showing functions in immune response and inflammation (Demetriou et al., 2001; Dong & Hughes, 1997; Dumic et al., 2006; Hughes, 2001). Other cell surface markers involved in cell-adhesion processes such as CD44 are bound by galectin-8 in a glycan-dependent manner underlining the importance of galectin-8 as matricellular protein involved in the regulation of cell-adhesion (Sebban et al., 2007).

All three galectins mentioned in this review are able to bind different integrin subunits. All bind to β1-integrins (Dumic et al., 2006; Furtak et al., 2001; Hughes, 2001; Sakaguchi et al., 2010; Zick et al., 2004). In this context galectin-3 binding to β1-integrins leads to an internalisation signal, regulating receptor amount on the cell surface and thereby influencing cell signalling aspects (Furtak et al., 2001). Other integrins such as αvβ3 integrin on endothelial cells or the αM subunit on macrophages are also bound by galectin-3 (Dong & Hughes, 1997; Markowska et al., 2010). Galectin-8 is known to have a major function in integrin-binding and integrin-mediated signalling (Zick et al., 2004). The binding of galectin-8 N-CRD to the β1-integrin-sunbunit is especially good as high affinity α2-3-

sialylated ligands are presented on this subunit (Diskin et al., 2009). Beside the β1-sunbunit galectin-8 N-CRD also binds α5 and some other integrin-subunits, but literature does not give a clear picture about the exact integrin binding partners. For example N-glycans on the α4-subunit are once mentioned as main binding partner while other authors do not report binding to this subunit. Similar discrepancies were noticed for other subunits (Cárcamo et al., 2006; Diskin et al., 2009; Hadari et al., 2000; Nishi et al., 2003; Yamamoto et al., 2008). This might be explained by tissue- or cell-specific glycosylation patterns of the single subunits. In contrast to most interactions which are performed by the N-terminal galectin-8 CRD binding to the αM-subunit is performed by the C-terminal CRD (Nishi et al., 2003).

5. Galectin function in cell adhesion and cell migration

5.1 Principle function

Galectins can act pro- or antiadhesive for different cell types. They can either facilitate or reduce adhesion to other cells depending on different factors. Cell adhesion is enhanced if galectins crosslink glycosylated structures on one cell with glycans on other cells or the extracellular matrix. In contrast the adhesion is reduced if soluble galectins block available receptors for other binding interactions. This depends on one hand on galectin concentration. At high concentrations galectins may block all available receptors without interaction with each other which is necessary for crosslinking and therefore for adhesion (Elola et al., 2007). It is for example discussed that galectin-3 outbursts can lead to detachment of cells from the extracellular matrix as galectin-3 blocks integrin binding to ECM glycoproteins (Ochieng et al., 1998b). On the other hand it is important which receptors are available on the specific cell type used in the experiment and if those receptors interact more easily with the soluble galectins or with receptors on the surface the cell attaches to (Elola et al., 2007). Additional the oligomerisation state of the galectins plays an important role as they can either block receptors or crosslink molecules depending on their valency (Hughes, 2001). The oligomerisation is in case of galectin-1 depending on galectin concentration while galectin-3 stays monomeric in solution without ligand binding and builds pentamers after the binding reaction (Ahmad et al., 2004a; Cho & Cummings, 1995; Cho & Cummings, 1997; Morris et al., 2004; Nieminen et al., 2008). Moreover effects of single galectins can hardly be determined as most cell types co-express different galectins which might at least partially result in overlapping or opposite effects (Cooper & Baronoes, 1999; Liu & Rabinovich, 2005).

In addition to direct binding of galectins to glycan structures on either membrane-bound receptors or ECM-glycoproteins, regulation of integrin amount, availability and affinity by galectin binding also contributes to adhesion events. Galectin-3 for example is able to increase amount and/or affinity of β2-integrins on the cell surface on neutrophils, thereby regulating the binding to ECM glycoproteins recognised by integrins (Hughes, 2001; Kuwabara & Liu, 1996). Overexpression of galectin-3 correlates with enhanced surface expression of α4β7 integrins resulting in increased cell adhesion (Matarrese et al., 2000). In contrast binding of galectin-3 leads to internalisation of β1-integrins in breast carcinoma cells (Furtak et al., 2001). Moreover the clustering and residence time of other receptors on the cell surface is regulated by the formation of glycan-galectin lattices thereby regulating different signalling processes (Garner & Baum, 2008; Lau & Dennis, 2008; Rabinovich et al., 2007).

5.2 Selected examples of cell adhesion and motility regulated by galectins-1, -3 and -8

We here present only few examples of galectin function in cell adhesion and motility processes. The list is by far not complete. Other review articles focus more detailed on cell adhesion events mediated by galectins (Elola et al., 2007; Hughes, 2001).

Galectin-1 is an important factor for the adhesion and proliferation of neural stem cells and neural progenitor cells. The adhesion is mediated by the carbohydrate recognition domain and interaction of this binding domain with integrin β1 subunit (Sakaguchi et al., 2006; Sakaguchi et al., 2010). Moreover galectin-1 can reduce the motility of immune cells which might explain parts of its anti-inflammatory effects (Elola et al., 2007; Liu, 2005; Rabinovich et al., 2002a; Rabinovich et al., 2002b).

One important function of galectin-3 is associated with angiogenesis (Nangia-Makker et al., 2000a; Nangia-Makker et al., 2000b). Galectin-3 increases for example angiogenesis by forming integrin αvβ3 lattices on the cell-surface leading to FAK regulated downstream signalling. Galectin-3 mediated angiogenesis depends on the growth factors VEGF and bFGF (Markowska et al., 2010). Another interesting function of galectin-3 is the chemotattraction of monocytes via a G-protein coupled receptor pathway and the role in eosinophil rolling to sites of inflammation (Rao et al., 2007; Sano et al., 2000). Most of those functions can only be performed by full length galectin-3 showing the importance of glycan binding and oligomerisation of the protein (Markowska et al., 2010; Sano et al., 2000). Different other biological activities are also depending on both N- and C-terminal domain (Nieminen et al., 2005; Ochieng et al., 1998a; Sano et al., 2000; Sato et al., 2002; Yamaoka et al., 1995). This proves the possibility of regulating galectin-3 function by protease cleavage as mentioned in chapter 2.3.3.

Galectin-8 has been assigned to matricellular proteins which are able to promote cell adhesion. CHO-cells on galectin-8 show similar binding kinetics as on fibronectin but differ in their formation of cytoskeleton (Boura-Halfon et al., 2003). Moreover the binding to galectin-8 triggers specific signalling cascades as Ras, MAPK and Erk pathway (Levy et al., 2003). A physiological function in human might be the modulation of neutrophil function. Galectin-8 promotes neutrophil adhesion by binding αM integrin and promatrix metalloproteinase-9. Moreover superoxide production which is essentiell for neutrophil function is triggered by galectin-8 C-terminal CRD (Nishi et al., 2003). Another galectin-8 function might be the promotion of angiogenesis as it was also shown for galectin-3. Galectin-8 increases tube formation *in vitro* and angiogenesis *in vivo* in dependence of its specific carbohydrate affinity at physiological concentrations. This regulatory function is at least partially depending on CD 166 (Delgado et al., 2011).

6. Galectins in biomaterial research

As discussed in chapter 5.1 galectins can act pro- and antiadhesive which *in vivo* seems to be mainly regulated by concentration and oligomerisation status of the galectins. In the context of biomaterial research it is also of huge importance if the galectins are immobilised or soluble presented. Immobilised galectins act mainly proadhesive as they crosslink the surface they are immobilised on with glycosylated structures on the cell-membrane. Soluble galectins can either facilitate or reduce adhesion for example to functionalised surfaces as discussed for the *in vivo* situation in chapter 5.1 depending on concentration, oligomerisation and cell type (respectively receptor availability on this cell type) (Elola et al., 2007).

The pro-adhesive properties of galectins have been shown several times. But only few efforts have been done to elucidate the potential of galectins as coatings for biomaterial surfaces. In contrast other components of the extracellular matrix are often used. Coatings with peptides from ECM proteins such as RGD or YIGSR peptide are one of the most common methods to modify biomaterial surfaces. Also coatings with complete ECM proteins or specific adhesion proteins have been investigated. Another important molecule class used in biomaterial research today are growth factors (Chan & Mooney, 2008; Shekaran & Garcia, 2011; Straley et al., 2010). The functionalisation with glycans or lectins seems to be underrepresented although their function in natural processes is well known. Only few studies show the potential of galectins and glycans as biomaterial coatings:

The positive influence of galectins was shown for example as the coating of PLGA scaffolds with recombinant galectin-1 promotes adhesion and growth of immortal rat chondrocytes. Therefore this surface is mentioned to have potential as biomaterial in tissue engineering (Chen et al., 2005). The potential of glycans in biomaterial coatings has also been shown. For example galactose derivatives immobilised on material surfaces were proven to influence the growth and function of liver cells positively. But in this study the receptor molecules and mechanisms of signal transduction were not investigated and binding of an asialoglycoprotein receptor (and not galectin mediated binding) is assumed (De Bartolo et al., 2006). Another study shows combined use of immobilised glycans with galectins as it evidences positive effects of endogenous galectin-1 for adhesion of chondrocytes to a lactose-modified surface (Marcon et al., 2005). These findings prove the possible use of glycan and/or galectin modified materials for improved cell adhesion.

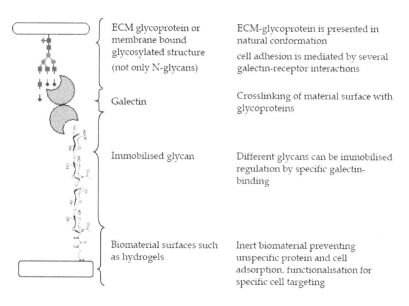

Fig. 4. Schematic representation of a possible biomaterial set-up using immobilised glycans as scaffold for subsequent galectin-mediated protein and cell-adhesion

Our recent work shows the potential of galectins and glycans in the preparation of biomaterial surfaces (figure 4). The assembly of an artificial extracellular matrix consisting of immobilised glycans, galectins and other extracellular matrix components was proven with a fungal model lectin (CGL2) (Sauerzapfe et al., 2009). In this approach poly-*N*-acetyllactosamine structures which are well known ligands for galectins (see chapter 3.2) are enzymatically produced. Those structures can be immobilised to different functionalised materials by a free amino group coupled to the reducing sugar. Concentration dependent binding of lectins to immobilised glycans was proven showing differences for specific glycan ligands. Lectin-mediated crosslinking of the surface with ECM-glycoproteins was also shown (Sauerzapfe et al., 2009). This galectin-mediated binding of ECM-glycoproteins leads to a natural presentation of these structures for subsequent adhesion of cells.

Our ongoing work focuses on the transfer of this set-up to applicable biomaterial surfaces. On the one hand recombinant human galectins are used instead of the fungal lectin to provide a more natural set-up (unpublished data). On the other hand the assembly of this artificial extracellular matrix is transferred to a special hydrogel surface. Star shaped NCO-sP(EO-stat-PO) is used as inert biomaterial which prevents unspecific protein adsorption and can be further functionalised with specific structures such as sugar molecules (Bruellhoff et al., 2010; Grafahrend et al., 2011). On the basis of these glycans an artificial extracellular matrix composed of galectins and ECM-glycoproteins can be built up. Fibroblasts show excellent adhesion and cell spreading on these surfaces (Rech, Beer, Elling, Groll, manuscript in preparation).

7. Conclusion

The importance of galectins in cell adhesion and signal transduction has been shown in several investigations. Therefore a possible application of galectins in the assembly of a biomaterial surface mimicking the natural microenvironment seems to be obvious. Anyhow only few articles regarding the use of galectins in biomaterial research have been published. This might be explained by the fact that the fine regulation of galectin mediated cell adhesion and signalling is still not fully understood yet. Therefore it is important to evaluate galectin function under specified conditions to reduce or exclude the risk of unwanted inflammatory or carcinogenic effects.

Taking the presented literature and our own work regarding the biofunctionalisation of surfaces with glycans and galectins together, there is clear evidence that galectins play an important role in cell adhesion and proliferation on specifically functionalised biomaterial surfaces. However, further research has to be done to adopt the fundamental understanding of galectin-glycan mediated cell adhesion processes to an applicable biomaterial surface.

8. Acknowledgements

The authors acknowledge financial support by the DFG (Deutsche Forschungsgemeinschaft) within the Research Training Group 1035 "Biointerface", by the DFG (project EL 135/10-1), and by the excellence initiative of the German federal and state governments through ERS@RWTH Aachen University.

9. References

Abbott, W. M. & Feizi, T. (1991). Soluble 14-kDa beta-galactoside-specific bovine lectin - evidence from mutagenesis and proteolysis that almost the complete polypeptide-chain is necessary for integrity of the carbohydrate recognition domain. *The Journal of Biological Chemistry*, 266, pp. 5552-5557.

Adams, J. C. (2001). Thrombospondins: Multifunctional Regulators of Cell Interactions. *Annual Review of Cell and Developmental Biology*, 17, pp. 25-51.

Ahmad, N., Gabius, H. J., Andre, S., Kaltner, H., Sabesan, S., Roy, R., Liu, B. C., Macaluso, F. & Brewer, C. F. (2004a). Galectin-3 precipitates as a pentamer with synthetic multivalent carbohydrates and forms heterogeneous cross-linked complexes. *The Journal of Biological Chemistry*, 279, pp. 10841-10847.

Ahmad, N., Gabius, H. J., Sabesan, S., Oscarson, S. & Brewer, C. F. (2004b). Thermodynamic binding studies of bivalent oligosaccharides to galectin-1, galectin-3, and the carbohydrate recognition domain of galectin-3. *Glycobiology*, 14, pp. 817-25.

Al-Ansari, S., Zeebregts, C. J., Slart, R., Peppelenbosch, M. & Tio, R. A. (2009). Galectins in atherosclerotic disease. *Trends in Cardiovascular Medicine*, 19, pp. 164-169.

Almkvist, J. & Karlsson, A. (2004). Galectins as inflammatory mediators. *Glycoconjugate Journal*, 19, pp. 575-581.

Appukuttan, P. S. (2002). Terminal alpha-linked galactose rather than N-acetyl lactosamine is ligand for bovine heart galectin-1 in N-linked oligosaccharides of glycoproteins. *Journal of Molecular Recognition*, 15, pp. 180-187.

Arumugham, R. G., Hsieh, T. C., Tanzer, M. L. & Laine, R. A. (1986). Structures of the asparagine-linked sugar chains of laminin. *Biochimica et Biophysica Acta*, 883, pp. 112-126.

Bachhawat-Sikder, K., Thomas, C. J. & Surolia, A. (2001). Thermodynamic analysis of the binding of galactose and poly-N-acetyllactosamine derivatives to human galectin-3. *FEBS Letters*, 500, pp. 75-79.

Barondes, S. H., Castronovo, V., Cooper, D. N. W., Cummings, R. D., Drickamer, K., Felzi, T., Gitt, M. A., Hirabayashi, J., Hughes, C., Kasai, K.-i., Leffler, H., Liu, F.-T., Lotan, R., Mercurio, A. M., Monsigny, M., Pillai, S., Poirer, F., Raz, A., Rigby, P. W. J., Rini, J. M. & Wang, J. L. (1994a). Galectins: A family of animal β-galactoside-binding lectins. *Cell*, 76, pp. 597-598.

Barondes, S. H., Cooper, D. N. W., Gitt, M. A. & Leffler, H. (1994b). Galectins - structure and function of a large family of animal lectins. *The Journal of Biological Chemistry*, 269, pp. 20807-20810.

Bidon, N., Brichory, F., Bourguet, P., Le Pennec, J. P. & Dazord, L. (2001). Galectin-8: A complex sub-family of galectins. *International Journal of Molecular Medicine*, 8, pp. 245-250.

Birdsall, B., Feeney, J., Burdett, I. D. J., Bawumia, S., Barboni, E. A. M. & Hughes, R. C. (2001). NMR solution studies of hamster galectin-3 and electron microscopic visualization of surface-adsorbed complexes: Evidence for interactions between the N- and C-terminal domains. *Biochemistry*, 40, pp. 4859-4866.

Blixt, O., Head, S., Mondala, T., Scanlan, C., Huflejt, M. E., Alvarez, R., Bryan, M. C., Fazio, F., Calarese, D., Stevens, J., Razi, N., Stevens, D. J., Skehel, J. J., van Die, I., Burton, D. R., Wilson, I. A., Cummings, R., Bovin, N., Wong, C. H. & Paulson, J. C. (2004). Printed covalent glycan array for ligand profiling of diverse glycan binding proteins. *Proceedings of the National Academy of Sciences of the United States of America*, 101, pp. 17033-17038.

Bohorov, O., Andersson-Sand, H., Hoffmann, J. & Blixt, O. (2006). Arraying glycomics: a novel bi-functional spacer for one-step microscale derivatization of free reducing glycans. *Glycobiology*, 16, pp. 21C-27C.

Boura-Halfon, S., Voliovitch, H., Feinstein, R., Paz, K. & Zick, Y. (2003). Extracellular matrix proteins modulate endocytosis of the insulin receptor. *The Journal of Biological Chemistry*, 278, pp. 16397-16404.

Brewer, C. F. (2004). Thermodynamic binding studies of galectin-1, -3 and -7. *Glycoconjugate Journal*, 19, pp. 459-465.

Bruellhoff, K., Fiedler, J., Moller, M., Groll, J. & Brenner, R. E. (2010). Surface coating strategies to prevent biofilm formation on implant surfaces. *International Journal of Artificial Organs*, 33, pp. 646-653.

Bunkenborg, J., Pilch, B. J., Podtelejnikov, A. V. & Wiśniewski, J. R. (2004). Screening for N-glycosylated proteins by liquid chromatography mass spectrometry. *PROTEOMICS*, 4, pp. 454-465.

Cárcamo, C., Pardo, E., Oyanadel, C., Bravo-Zehnder, M., Bull, P., Cáceres, M., Martínez, J., Massardo, L., Jacobelli, S., González, A. & Soza, A. (2006). Galectin-8 binds specific beta1 integrins and induces polarized spreading highlighted by asymmetric lamellipodia in Jurkat T cells. *Experimental Cell Research*, 312, pp. 374-386.

Carlsson, S., Oberg, C. T., Carlsson, M. C., Sundin, A., Niisson, U. J., Smith, D., Cummings, R. D., Almkvist, J., Karlsson, A. & Leffler, H. (2007). Affinity of galectin-8 and its carbohydrate recognition domains for ligands in solution and at the cell surface. *Glycobiology*, 17, pp. 663-676.

Chan, G. & Mooney, D. J. (2008). New materials for tissue engineering: towards greater control over the biological response. *Trends in Biotechnology*, 26, pp. 382-392.

Chen, R., Jiang, X., Sun, D., Han, G., Wang, F., Ye, M., Wang, L. & Zou, H. (2009). Glycoproteomics analysis of human liver tissue by combination of multiple enzyme digestion and hydrazide chemistry. *Journal of Proteome Research*, 8, pp. 651-661.

Chen, S. J., Lin, C. C., Tuan, W. C., Tseng, C. S. & Huang, R. N. (2005). Effect of recombinant galectin-1 on the growth of immortal rat chondrocyte on chitosan-coated PLGA scaffold. *Journal of Biomedical Materials Research Part A*, 93A, pp. 1482-1492.

Chenna, R., Sugawara, H., Koike, T., Lopez, R., Gibson, T. J., Higgins, D. G. & Thompson, J. D. (2003). Multiple sequence alignment with the Clustal series of programs. *Nucleic Acids Research*, 31, pp. 3497-3500.

Cho, M. & Cummings, R. D. (1995). Galectin-1, a beta-galactoside-binding lectin in Chinese hamster ovary cells. I. Physical and chemical characterization. *The Journal of Biological Chemistry*, 270, pp. 5198-5206.

Cho, M. J. & Cummings, R. D. (1997). Galectin-1: Oligomeric structure and interactions with polylactosamine. *Trends in Glycoscience and Glycotechnology*, 9, pp. 47-56.

Cooper, D. N. W. (1997). Galectin-1: Secretion and modulation of cell interactions with laminin. *Trends in Glycoscience and Glycotechnology*, 9, pp. 57-67.

Cooper, D. N. W. & Barondes, S. H. (1999). God must love galectins; He made so many of them. *Glycobiology*, 9, pp. 979-984.

Cooper, D. N. W. (2002). Galectinomics: finding themes in complexity. *Biochimica et Biophysica Acta - General Subjects*, 1572, pp. 209-231.

Dam, T. K., Gabius, H. J., Andre, S., Kaltner, H., Lensch, M. & Brewer, C. F. (2005). Galectins bind to the multivalent glycoprotein asialofetuin with enhanced affinities and a gradient of decreasing binding constants. *Biochemistry*, 44, pp. 12564-12571.

Danguy, A., Camby, I. & Kiss, R. (2002). Galectins and cancer. *Biochimica et Biophysica Acta - General Subjects*, 1572, pp. 285-293.

De Bartolo, L., Morelli, S., Rende, M., Salerno, S., Giorno, L., Lopez, L. C., Favia, P., d'Agostino, R. & Drioli, E. (2006). Galactose derivative immobilized glow discharge processed polyethersulfone membranes maintain the liver cell metabolic activity. *Journal of Nanoscience and Nanotechnology*, 6, pp. 2344-2353.

Delgado, V. M. C., Nugnes, L. G., Colombo, L. L., Troncoso, M. F., Fernandez, M. M., Malchiodi, E. L., Frahm, I., Croci, D. O., Compagno, D., Rabinovich, G. A., Wolfenstein-Todel, C. & Elola, M. T. (2011). Modulation of endothelial cell migration and angiogenesis: a novel function for the "tandem-repeat" lectin galectin-8. *The FASEB Journal*, 25, pp. 242-254.

Dell, A. (2002). Structures of Glycoprotein Glycans. *Australian Journal of Chemistry*, 55, pp. 27-37.

Demetriou, M., Granovsky, M., Quaggin, S. & Dennis, J. W. (2001). Negative regulation of T-cell activation and autoimmunity by Mgat5 N-glycosylation. *Nature*, 409, pp. 733-739.

Di Lella, S., Ma, L., Díaz Ricci, J. C., Rabinovich, G. A., Asher, S. A. & Álvarez, R. M. S. (2009). Critical role of the solvent environment in galectin-1 binding to the disaccharide lactose. *Biochemistry*, 48, pp. 786-791.

Di Virgilio, S., Glushka, J., Moremen, K. & Pierce, M. (1999). Enzymatic synthesis of natural and C-13 enriched linear poly-N-acetyllactosamines as ligands for galectin-1. *Glycobiology*, 9, pp. 353-364.

Diehl, C., Engstrom, O., Delaine, T., Hakansson, M., Genheden, S., Modig, K., Leffler, H., Ryde, U., Nilsson, U. J. & Akke, M. (2010). Protein flexibility and conformational entropy in ligand design targeting the carbohydrate recognition domain of galectin-3. *Journal of the American Chemical Society*, 132, pp. 14577-14589.

Diskin, S., Cao, Z. Y., Leffler, H. & Panjwani, N. (2009). The role of integrin glycosylation in galectin-8-mediated trabecular meshwork cell adhesion and spreading. *Glycobiology*, 19, pp. 29-37.

Do, K.-Y., Smith, D. F. & Cummings, R. D. (1990). LAMP-1 in cho cells is a primary carrier of poly-N-acetyllactosamine chains and is bound preferentially by a mammalian S-type lectin. *Biochemical and Biophysical Research Communications*, 173, pp. 1123-1128.

Dong, S. & Hughes, R. C. (1997). Macrophage surface glycoproteins binding to galectin-3 (Mac-2-antigen). *Glycoconjugate Journal*, 14, pp. 267-274.

Dumic, J., Dabelic, S. & Flogel, M. (2006). Galectin-3: An open-ended story. *Biochimica Et Biophysica Acta-General Subjects*, 1760, pp. 616-635.

Elola, M. T., Wolfenstein-Todel, C., Troncoso, M. F., Vasta, G. R. & Rabinovich, G. A. (2007). Galectins: matricellular glycan-binding proteins linking cell adhesion, migration, and survival. *Cellular and Molecular Life Sciences*, 64, pp. 1679-1700.

Furtak, V., Hatcher, F. & Ochieng, J. (2001). Galectin-3 mediates the endocytosis of beta-1 integrins by breast carcinoma cells. *Biochemical and Biophysical Research Communications*, 289, pp. 845-850.

Garner, O. B. & Baum, L. G. (2008). Galectin-glycan lattices regulate cell-surface glycoprotein organization and signalling. *Biochemical Society Transactions*, 36, pp. 1472-1477.

Giguere, D., Bonin, M. A., Cloutier, P., Patnam, R., St-Pierre, C., Sato, S. & Roy, R. (2008). Synthesis of stable and selective inhibitors of human galectins-1 and-3. *Bioorganic & Medicinal Chemistry*, 16, pp. 7811-7823.

Gong, H. C., Honjo, Y., Nangia-Makker, P., Hogan, V., Mazurak, N., Bresalier, R. S. & Raz, A. (1999). The NH$_2$ terminus of galectin-3 governs cellular compartmentalization and functions in cancer cells. *Cancer Research*, 59, pp. 6239-6245.

Grafahrend, D., Heffels, K. H., Beer, M. V., Gasteier, P., Moller, M., Boehm, G., Dalton, P. D. & Groll, J. (2011). Degradable polyester scaffolds with controlled surface chemistry combining minimal protein adsorption with specific bioactivation. *Nature Materials*, 10, pp. 67-73.

Gu, M., Wang, W., Song, W. K., Cooper, D. N. W. & Kaufman, S. J. (1994). Selective modulation of the interaction of alpha7beta1 integrin with fibronectin and laminin by L-14 lectin during skeletal muscle differentiation. *Journal of Cell Science*, 107, pp. 175-181.

Guévremont, M., Martel-Pelletier, Boileau, C., Liu, F.-T., Richard, M., Fernandes, J.-C., Pelletier, J.-P. & Reboul, P. (2004). Galectin-3 surface expression on human adult chondrocytes: a potential substrate for collagenase-3. *Annals of the Rheumatic Diseases*, pp. 636-643.

Guex, N. & Peitsch, M. C. (1997). Swiss-Model and the swiss-pdb viewer: an environment for comparative protein modeling. *Electrophoresis*, 18, pp. 2714-2723.

Hadari, Y. R., Arbel-Goren, R., Levy, Y., Amsterdam, A., Alon, R., Zakut, R. & Zick, Y. (2000). Galectin-8 binding to integrins inhibits cell adhesion and induces apoptosis. *Journal of Cell Science*, 113, pp. 2385-2397.

Hernandez, J. D. & Baum, L. G. (2002). Ah, sweet mystery of death! Galectins and control of cell fate. *Glycobiology*, 12, pp. 127R-136.

Hirabayashi, J. & Kasai, K. (1991). Effect of amino acid substitution by sited-directed mutagenesis on the carbohydrate recognition and stability of human 14-kDa beta-galactoside-binding lectin. *The Journal of Biological Chemistry*, 266, pp. 23648-23653.

Hirabayashi, J. & Kasai, K. (1993). The family of metazoan metal-independent beta-galactoside-binding lectins: structure, function and molecular evolution. *Glycobiology*, 3, pp. 297-304.

Hirabayashi, J. & Kasai, K. I. (1994). Further evidence by site-directed mutagenesis that conserved hydrophilic residues form a carbohydrate-binding site of human galectin-1. *Glycoconjugate Journal*, 11, pp. 437-442.

Hirabayashi, J., Hashidate, T., Arata, Y., Nishi, N., Nakamura, T., Hirashima, M., Urashima, T., Oka, T., Futai, M., Muller, W. E. G., Yagi, F. & Kasai, K.-i. (2002). Oligosaccharide specificity of galectins: a search by frontal affinity chromatography. *Biochimica et Biophysica Acta - General Subjects*, 1572, pp. 232-254.

Horie, H., Kadoya, T., Hikawa, N., Sango, K., Inoue, H., Takeshita, K., Asawa, R., Hiroi, T., Sato, M., Yoshioka, T. & Ishikawa, Y. (2004). Oxidized galectin-1 stimulates macrophages to promote axonal regeneration in peripheral nerves after axotomy. *The Journal of Neuroscience*, 24, pp. 1873-1880.

Houzelstein, D., Gonc‚alves, I. R., Fadden, A. J., Sidhu, S. S., Cooper, D. N. W., Drickamer, K., Leffler, H. & Poirier, F. (2004). Phylogenetic analysis of the vertebrate galectin family. *Molecular Biology and Evolution*, 21, pp. 1177–1187.

Hughes, R. C. (1999). Secretion of the galectin family of mammalian carbohydrate-binding proteins. *Biochimica et Biophysica Acta*, 1473, pp. 172-185.

Hughes, R. C. (2001). Galectins as modulators of cell adhesion. *Biochimie*, 83, pp. 667-676.

Ideo, H., Seko, A., Ishizuka, I. & Yamashita, K. (2003). The N-terminal carbohydrate recognition domain of galectin-8 recognizes specific glycosphingolipids with high affinity. *Glycobiology*, 13, pp. 713-723.

Ideo, H., Matsuzaka, T., Nonaka, T., Seko, A. & Yamashita, K. (2011). Galectin-8-N-domain recognition mechanism for sialylated and sulfated glycans. *The Journal of Biological Chemistry*, pp.11346-11355

Ilarregui, J. M., Bianco, G. A., Toscano, M. A. & Rabinovich, G. A. (2005). The coming of age of galectins as immunomodulatory agents: impact of these carbohydrate binding proteins in T cell physiology and chronic inflammatory disorders. *Annals of the Rheumatic Diseases*, 64, pp. 96-103.

Janik, M. E., Litynska, A. & Vereecken, P. (2010). Cell migration-The role of integrin glycosylation. *Biochimica et Biophysica Acta- General Subjects*, 1800, pp. 545-555.

John, C. M., Leffler, H., Kahl-Knutsson, B., Svensson, I. & Jarvis, G. A. (2003). Truncated galectin-3 inhibits tumor growth and metastasis in orthotopic nude mouse model of human breast cancer. *Clinical Cancer Research*, 9, pp. 2374-2383.

Kariya, Y., Kato, R., Itoh, S., Fukuda, T., Shibukawa, Y., Sanzen, N., Sekiguchi, K., Wada, Y., Kawasaki, N. & Gu, J. G. (2008). N-Glycosylation of laminin-332 regulates its biological functions: A novel function of the bisecting GlcNAc. *The Journal of Biological Chemistry*, 283, pp. 33036-33045.

Kishishita, S., Nishino, A., Murayama, K., Terada, T., Shirouzu, M. & Yokoyama, S. (2008). pdb database 2YXS.

Klyosov, A. A., Zbigniew, J. W. & Platt, D. (2008). *Galectins*. Wiley, ISBN 978-0-470-37318, Hoboken.

Knibbs, R. N., Perini, F. & Goldstein, I. J. (1989). Structure of the Major Concanavalin-a Reactive Oligosaccharides of the Extracellular-Matrix Component Laminin. *Biochemistry*, 28, pp. 6379-6392.

Knibbs, R. N., Agrwal, N., Wang, J. L. & Goldstein, I. J. (1993). Carbohydrate-binding protein-35 2. Analysis of the interaction of the recombinant polypeptide with saccharides. *The Journal of Biological Chemistry*, 268, pp. 14940-14947.

Kubler, D., Hung, C. W., Dam, T. K., Kopitz, J., Andre, S., Kaltner, H., Lohr, M., Manning, J. C., He, L. Z., Wang, H., Middelberg, A., Brewer, C. F., Reed, J., Lehmann, W. D. & Gabius, H. J. (2008). Phosphorylated human galectin-3: Facile large-scale preparation of active lectin and detection of structural changes by CD spectroscopy. *Biochimica et Biophysica Acta- General Subjects*, 1780, pp. 716-722.

Kuwabara, I. & Liu, F. T. (1996). Galectin-3 promotes adhesion of human neutrophils to laminin. *Journal of Immunology*, 156, pp. 3939-3944.

Larkin, M. A., Blackshields, G., Brown, N. P., Chenna, R., McGettigan, P. A., McWilliam, H., Valentin, F., Wallace, I. M., Wilm, A., Lopez, R., Thompson, J. D., Gibson, T. J. & Higgins, D. G. (2007). Clustal W and clustal X version 2.0. *Bioinformatics*, 23, pp. 2947-2948.

Lau, K. S. & Dennis, J. W. (2008). N-Glycans in cancer progression. *Glycobiology*, 18, pp. 750-760.

Leffler, H., Carlsson, S., Hedlund, M., Qian, Y. & Poirier, F. (2004). Introduction to galectins. *Glycoconjugate Journal*, 19, pp. 433-440.

Leppänen, A., Stowell, S., Blixt, O. & Cummings, R. D. (2005). Dimeric galectin-1 binds with high affinity to alpha2,3-sialylated and non-sialylated terminal N-Acetyllactosamine units on surface-bound extended glycans. *The Journal of Biological Chemistry*, 280, pp. 5549-5562.

Levy, Y., Ronen, D., Bershadsky, A. D. & Zick, Y. (2003). Sustained induction of ERK, protein kinase B, and p70 S6 kinase regulates cell spreading and formation of F-actin microspikes upon ligation of integrins by galectin-8, a mammalian lectin. *The Journal of Biological Chemistry*, 278, pp. 14533-14542.

Levy, Y., Auslender, S., Eisenstein, M., Vidavski, R. R., Ronen, D., Bershadsky, A. D. & Zick, Y. (2006). It depends on the hinge: a structure-functional analysis of galectin-8, a tandem-repeat type lectin. *Glycobiology*, 16, pp. 463-476.

Liu, F.-T., Patterson, R. J. & Wang, J. L. (2002). Intracellular functions of galectins. *Biochimica et Biophysica Acta (BBA) - General Subjects*, 1572, pp. 263-273.

Liu, F.-T. (2005). Regulatory Roles of Galectins in the Immune Response. *International Archives of Allergy and Immunology*, 136, pp. 385-400.

Liu, F.-T. & Rabinovich, G. A. (2005). Galectins as modulators of tumour progression. *Nature Reviews Cancer*, 5, pp. 29-41.

Liu, T., Qian, W.-J., Gritsenko, M. A., Camp, D. G., Monroe, M. E., Moore, R. J. & Smith, R. D. (2005). Human Plasma N-Glycoproteome Analysis by Immunoaffinity Subtraction, Hydrazide Chemistry, and Mass Spectrometry. *Journal of Proteome Research*, 4, pp. 2070-2080.

Lobsanov, Y. D., Gitt, M. A., Leffler, H., Barondes, S. H. & Rini, J. M. (1993). X-ray crystal structure of the human dimeric S-Lac lectin, L-14-II, in complex with lactose at 2.9-A resolution. *The Journal of Biological Chemistry*, 268, pp. 27034-27038.

Lopez-Lucendo, M. F., Solis, D., Andre, S., Hirabayashi, J., Kasai, K., Kaltner, H., Gabius, H. J. & Romero, A. (2004). Growth-regulatory human galectin-1: crystallographic characterisation of the structural changes induced by single-site mutations and their impact on the thermodynamics of ligand binding. *The Journal of Molecular Biology*, 343, pp. 957-970.

Marcon, P., Marsich, E., Vetere, A., Mozetic, P., Campa, C., Donati, I., Vittur, F., Gamini, A. & Paoletti, S. (2005). The role of Galectin-1 in the interaction between chondrocytes and a lactose-modified chitosan. *Biomaterials*, 26, pp. 4975-4984.

Markowska, A. I., Liu, F. T. & Panjwani, N. (2010). Galectin-3 is an important mediator of VEGF- and bFGF-mediated angiogenic response. *Journal of Experimental Medicine*, 207, pp. 1981-1993.

Massa, S. M., Cooper, D. N. W., Leffler, H. & Barondes, S. H. (1993). L-29, an endogenous lectin, binds to glycoconjugate ligands with positive cooperativity. *Biochemistry*, 32, pp. 260-267.

Masuda, K., Takahashi, N., Tsukamoto, Y., Honma, H. & Kohri, K. (2000). N-Glycan structures of an osteopontin from human bone. *Biochemical and Biophysical Research Communications*, 268, pp. 814-817.

Matarrese, P., Fusco, O., Tinari, N., Natoli, C., Liu, F. T., Semeraro, M. L., Malorni, W. & Iacobelli, S. (2000). Galectin-3 overexpression protects from apoptosis by improving cell adhesion properties. *International Journal of Cancer*, 85, pp. 545-554.

Mazurek, N., Conklin, J., Byrd, J. C., Raz, A. & Bresalier, R. S. (2000). Phosphorylation of the beta-galactoside-binding protein galectin-3 modulates binding to its ligands. *The Journal of Biological Chemistry*, 275, pp. 36311-36315.

Mehul, B. & Hughes, R. C. (1997). Plasma membrane targetting, vesicular budding and release of galectin 3 from the cytoplasm of mammalian cells during secretion. *Journal of Cell Science*, 110, pp. 1169-1178.

Moiseeva, E. P., Spring, E. L., Baron, J. H. & de Bono, D. P. (1999). Galectin 1 modulates attachment, spreading and migration of cultured vascular smooth muscle cells via interactions with cellular receptors and components of extracellular matrix. *Journal of Vascular Research*, 36, pp. 47-58.

Moiseeva, E. P., Williams, B. & Samani, N. J. (2003). Galectin 1 inhibits incorporation of vitronectin and chondroitin sulfate B into the extracellular matrix of human vascular smooth muscle cells. *Biochimica et Biophysica Acta*, 1619, pp. 125-132.

Morris, S., Ahmad, N., Andre, S., Kaltner, H., Gabius, H. J., Brenowitz, M. & Brewer, F. (2004). Quaternary solution structures of galectins-1, -3, and -7. *Glycobiology*, 14, pp. 293-300.

Munoz, F. J., Santos, J. I., Arda, A., Andre, S., Gabius, H. J., Sinisterra, J. V., Jimenez-Barbero, J. & Hernaiz, M. J. (2010). Binding studies of adhesion/growth-regulatory galectins with glycoconjugates monitored by surface plasmon resonance and NMR spectroscopy. *Organic & Biomolecular Chemistry*, 8, pp. 2986-2992.

Nangia-Makker, P., Baccarini, S. & Raz, A. (2000a). Carbohydrate-recognition and angiogenesis. *Cancer and Metastasis Reviews*, 19, pp. 51-57.

Nangia-Makker, P., Honjo, Y., Sarvis, R., Akahani, S., Hogan, V., Pienta, K. J. & Raz, A. (2000b). Galectin-3 induces endothetial cell morphogenesis and angiogenesis. *American Journal of Pathology*, 156, pp. 899-909.

Nangia-Makker, P., Balan, V. & Raz, A. (2008). Regulation of tumor progression by extracellular Galectin-3. *Cancer Microenvironment*, 1, pp.43-51

Nieminen, J., St-Pierre, C. & Sato, S. (2005). Galectin-3 interacts with naive and primed neutrophils, inducing innate immune responses. *Journal of Leukocyte Biology*, 78, pp. 1127-1135.

Nieminen, J., Kuno, A., Hirabayashi, J. & Sato, S. (2008). Visualization of Galectin-3 Oligomerization on the Surface of Neutrophils and Endothelial Cells Using Fluorescence Resonance Energy Transfer. *The Journal of Biological Chemistry*, 282, pp. 1374-1383.

Nishi, N., Shoji, H., Seki, M., Itoh, A., Miyanaka, H., Yuube, K., Hirashima, M. & Nakamura, T. (2003). Galectin-8 modulates neutrophil function via interaction with integrin alpha M. *Glycobiology*, 13, pp. 755-763.

Nishi, N., Itoh, A., Shoji, H., Miyanaka, H. & Nakamura, T. (2006). Galectin-8 and galectin-9 are novel substrates for thrombin. *Glycobiology*, 16, pp. 15C-20C.

Nishi, N., Abe, A., Iwaki, J., Yoshida, H., Itoh, A., Shoji, H., Kamitori, S., Hirabayashi, J. & Nakamura, T. (2008). Functional and structural bases of a cysteine-less mutant as a long-lasting substitute for galectin-1. *Glycobiology*, pp. 1065-1073.

Ochieng, J., Fridman, R., Nangiamakker, P., Kleiner, D. E., Liotta, L. A., Stetlerstevenson, W. G. & Raz, A. (1994). Galectin-3 is a novel substrate for human matrix metalloproteinase-2 and metalloproteinase-9. *Biochemistry*, 33, pp. 14109-14114.

Ochieng, J., Green, B., Evans, S., James, O. & Warfield, P. (1998a). Modulation of the biological functions of galectin-3 by matrix metalloproteinases. *Biochimica et Biophysica Acta- General Subjects*, 1379, pp. 97-106.

Ochieng, J., Leite-Browning, M. L. & Warfield, P. (1998b). Regulation of cellular adhesion to extracellular matrix proteins by galectin-3. *Biochemical and Biophysical Research Communications*, 246, pp. 788-791.

Ozeki, Y., Matsui, T., Yamamoto, Y., Funahashi, M., Hamako, J. & Titani, K. (1995). Tissue fibronectin is an endogenous ligand for galectin-1. *Glycobiology*, 5, pp. 255-261.

Pace, K. E., Lee, C., Stewart, P. L. & Baum, L. G. (1999). Restricted receptor segregation into membrane microdomains occurs on human T cells during apoptosis induced by galectin-1. *Journal of Immunology*, 163, pp. 3801-3811.

Patnaik, S. K., Potvin, B., Carlsson, S., Sturm, D., Leffler, H. & Stanley, P. (2006). Complex N-glycans are the major ligands for galectin-1, -3, and -8 on Chinese hamster ovary cells. *Glycobiology*, 16, pp. 305-317.

Paul, J. I. & Hynes, R. O. (1984). Multiple fibronectin subunits and their post-translational modifications. *The Journal of Biological Chemistry*, 259, pp. 13477-13487.

Probstmeier, R., Montag, D. & Schachner, M. (1995). Galectin-3, a beta-galactoside-binding animal lectin, binds to neural recognition molecules. *Journal of Neurochemistry*, 64, pp. 2465-2472.

Rabinovich, G. A., Baum, L. G., Tinari, N., Paganelli, R., Natoli, C., Liu, F. T. & Iacobelli, S. (2002a). Galectins and their ligands: amplifiers, silencers or tuners of the inflammatory response? *Trends in Immunology*, 23, pp. 313-320.

Rabinovich, G. A., Rubinstein, N. & Toscano, M. A. (2002b). Role of galectins in inflammatory and immunomodulatory processes. *Biochimica et Biophysica Acta-General Subjects*, 1572, pp. 274-284.

Rabinovich, G. A., Toscano, M. A., Jackson, S. S. & Vasta, G. R. (2007). Functions of cell surface galectin-glycoprotein lattices. *Current Opinion in Structural Biology*, 17, pp. 513-520.

Rabinovich, G. A. & Toscano, M. A. (2009). Turning 'sweet' on immunity: galectin-glycan interactions in immune tolerance and inflammation. *Nature Reviews Immunology*, 9, pp. 338-352.

Rao, S. P., Wang, Z. Z., Zuberi, R. I., Sikora, L., Bahaie, N. S., Zuraw, B. L., Liu, F. T. & Sriramarao, P. (2007). Galectin-3 functions as an adhesion molecule to support eosinophil rolling and adhesion under conditions of flow. *The Journal of Immunology*, 179, pp. 7800-7807.

Rapoport, E. M., Andre, S., Kurmyshkina, O. V., Pochechueva, T. V., Severov, V. V., Pazynina, G. V., Gabius, H.-J. & Bovin, N. V. (2008). Galectin-loaded cells as a platform for the profiling of lectin specificity by fluorescent neoglycoconjugates: A case study on galectins-1 and -3 and the impact of assay setting. *Glycobiology*, 18, pp. 315-324.

Sakaguchi, M., Shingo, T., Shimazaki, T., Okano, H. J., Shiwa, M., Ishibashi, S., Oguro, H., Ninomiya, M., Kadoya, T., Horie, H., Shibuya, A., Mizusawa, H., Poirier, F., Nakauchi, H., Sawamoto, K. & Okano, H. (2006). A carbohydrate-binding protein,

Galectin-1, promotes proliferation of adult neural stem cells. *Proceedings of the National Academy of Sciences of the United States of America*, 103, pp. 7112-7117.

Sakaguchi, M., Imaizumi, Y., Shingo, T., Tada, H., Hayama, K., Yamada, O., Morishita, T., Kadoya, T., Uchiyama, N., Shimazaki, T., Kuno, A., Poirier, F., Hirabayashi, J., Sawamoto, K. & Okano, H. (2010). Regulation of adult neural progenitor cells by Galectin-1/beta 1 Integrin interaction. *Journal of Neurochemistry*, 113, pp. 1516-1524.

Salameh, B. A., Cumpstey, I., Sundin, A., Leffler, H. & Nilsson, U. J. (2010). 1H-1,2,3-Triazol-1-yl thiodigalactoside derivatives as high affinity galectin-3 inhibitors. *Bioorganic & Medicinal Chemistry*, 18, pp. 5367-5378.

Salomonsson, E., Carlsson, M. C., Osla, V., Hendus-Altenburger, R., Kahl-Knutson, B., Oberg, C. T., Sundin, A., Nilsson, R., Nordberg-Karlsson, E., Nilsson, U. J., Karlsson, A., Rini, J. M. & Leffler, H. (2010). Mutational tuning of galectin-3 specificity and biological function. *The Journal of Biological Chemistry*, 285, pp. 35079-35091.

Sano, H., Hsu, D. K., Yu, L., Apgar, J. R., Kuwabara, I., Yamanaka, T., Hirashima, M. & Liu, F.-T. (2000). Human galectin-3 is a novel chemoattractant for monocytes and macrophages. *The Journal of Immunology*, 165, pp. 2156-2164.

Sasaki, T., Brakebusch, C., Engel, J. & Timpl, R. (1998). Mac-2 binding protein is a cell-adhesive protein of the extracellular matrix which self-assembles into ring-like structures and binds beta1 integrins, collagens and fibronectin. *The EMBO Journal*, 17, pp. 1606-1613.

Sato, S. & Hughes, R. C. (1992). Binding-specificity of a baby hamster-kidney lectin for H type-I and type-II chains, polylactosamine glycans, and appropriately glycosylated forms of laminin and fibronectin. *The Journal of Biological Chemistry*, 267, pp. 6983-6990.

Sato, S., Ouellet, N., Pelletier, I., Simard, M., Rancourt, A. & Bergeron, M. G. (2002). Role of galectin-3 as an adhesion molecule for neutrophil extravasation during streptococcal pneumonia. *Journal of Immunology*, 168, pp. 1813-1822.

Sauerzapfe, B., Krenek, K., Schmiedel, J., Wakarchuk, W. W., Pelantová, H., Kren, V. & Elling, L. (2009). Chemo-enzymatic synthesis of poly-*N*-acetyllactosamine (poly-LacNAc) structures and their characterization for CGL2-galectin-mediated binding of ECM glycoproteins to biomaterial surfaces. *Glycoconjugate Journal*, 26, pp. 141-159.

Sebban, L. E., Ronen, D., Levartovsky, D., Elkayam, O., Caspi, D., Aamar, S., Amital, H., Rubinow, A., Golan, I., Naor, D., Zick, Y. & Golan, I. (2007). The involvement of CD44 and its novel ligand galectin-8 in apoptotic regulation of autoimmune inflammation. *Journal of Immunology*, 179, pp. 1225-1235.

Seelenmeyer, C., Wegehingel, S., Tews, I., Kunzler, M., Aebi, M. & Nickel, W. (2005). Cell surface counter receptors are essential components of the unconventional export machinery of galectin-1. *The Journal of Cell Biology*, 171, pp. 373-381.

Seelenmeyer, C., Stegmayer, C. & Nickel, W. (2008). Unconventional secretion of fibroblast growth factor 2 and galectin-1 does not require shedding of plasma membrane-derived vesicles. *FEBS Letters*, 582, pp. 1362-1368.

Seetharaman, J., Kanigsberg, A., Slaaby, R., Leffler, H., Barondes, S. H. & Rini, J. R. (1998). X-ray crystal structure of the human galectin-3 carbohydrate recognition domain at 2.1-Å resolution. *The Journal of Biological Chemistry*, 273, pp. 13047-13052.

Shekaran, A. & Garcia, A. J. (2011). Extracellular matrix-mimetic adhesive biomaterials for bone repair. *Journal of Biomedical Materials Research Part A*, 96A, pp. 261-272.

Shekhar, M. P. V., Nangia-Makker, P., Tait, L., Miller, F. & Raz, A. (2004). Alterations in galectin-3 expression and distribution correlate with breast cancer progression - Functional analysis of galectin-3 in breast epithelial-endothefial interactions. *American Journal of Pathology*, 165, pp. 1931-1941.

Singh, P., Carraher, C. & Schwarzbauer, J. E. (2010). Assembly of fibronectin extracellular matrix. In: *Annual Review of Cell and Developmental Biology*, vol. 26, pp. 397-419, ISBN 1081-0706.

Song, X. Z., Lasanajak, Y., Xia, B. Y., Smith, D. & Cummings, R. (2009a). Fluorescent glycosylamides produced by microscale derivatization of free glycans for natural glycan microarrays. *ACS Chemical Biology*, vol. 4, no. 9, 741-750.

Song, X. Z., Xia, B. Y., Stowell, S. R., Lasanajak, Y., Smith, D. F. & Cummings, R. D. (2009b). Novel fluorescent glycan microarray strategy reveals ligands for galectins. *Chemistry & Biology*, 16, pp. 36-47.

Song, X. Z., Lasanajak, Y., Xia, B. Y., Heimburg-Molinaro, J., Rhea, J., Ju, H., Zhao, C. M., Molinaro, R., Cummings, R. & Smith, D. (2010). Shotgun glycomics: Functional identification of glycan determinants through a microarray strategy using fluorescently tagged glycans. *Glycobiology*, 20, pp. 54.

Sörme, P., Qian, Y. N., Nyholm, P. G., Leffler, H. & Nilsson, U. J. (2002). Low micromolar inhibitors of galectin-3 based on 3'-derivatization of N-acetyllactosamine. *Chembiochem*, 3, pp. 183-189.

Sörme, P., Kahl-Knutsson, B., Huflejt, M., Nilsson, U. J. & Leffler, H. (2004). Fluorescence polarization as an analytical tool to evaluate galectin-ligand interactions. *Analytical Biochemistry*, 334, pp. 36-47.

Sörme, P., Arnoux, P., Kahl-Knutsson, B., Leffler, H., Rini, J. M. & Nilsson, U. J. (2005). Structural and thermodynamic studies on cation-II interactions in lectin-ligand complexes: High-affinity galectin-3 inhibitors through fine-tuning of an arginine-arene interaction. *Journal of the American Chemical Society*, 127, pp. 1737-1743.

Stowell, S. R., Dias-Baruffi, M., Penttila, L., Renkonen, O., Nyame, A. K. & Cummings, R. D. (2004). Human galectin-1 recognition of poly-N-acetyllactosamine and chimeric polysaccharides. *Glycobiology*, 14, pp. 157-67.

Stowell, S. R., Arthur, C. M., Mehta, P., Slanina, K. A., Blixt, O., Leffler, H., Smith, D. F. & Cummings, R. D. (2008a). Galectin-1,-2, and-3 exhibit differential recognition of sialylated glycans and blood group antigens. *The Journal of Biological Chemistry*, 283, pp. 10109-10123.

Stowell, S. R., Arthur, C. M., Slanina, K. A., Horton, J. R., Smith, D. F. & Cummings, R. D. (2008b). Dimeric galectin-8 induces phosphatidylserine exposure in leukocytes through polylactosamine recognition by the C-terminal domain. *The Journal of Biological Chemistry*, 283, pp. 20547-20559.

Straley, K. S., Foo, C. W. P. & Heilshorn, S. C. (2010). Biomaterial Design Strategies for the Treatment of Spinal Cord Injuries. *Journal of Neurotrauma*, 27, pp. 1-19.

Szabo, P., Dam, T. K., Smetana, K., Dvorankova, B., Kurbler, D., Brewer, C. F. & Gabius, H. J. (2009). Phosphorylated Human Lectin Galectin-3: Analysis of Ligand Binding by Histochemical Monitoring of Normal/Malignant Squamous Epithelia and by Isothermal Titration Calorimetry. *Anatomia Histologia Embryologia*, 38, pp. 68-75.

Tanzer, M. L., Chandrasekaran, S., Dean, J. W., 3rd & Giniger, M. S. (1993). Role of laminin carbohydrates on cellular interactions. *Kidney International*, 43, pp. 66-72.

Tinari, N., Kuwabara, I., Huflejt, M. E., Shen, P. F., Iacobelli, S. & Liu, F.-T. (2001). Glycoprotein 90K/Mac-2BP interacts with galectin-1 and mediates galectin-1-induced cell aggregation. *International Journal of Cancer*, 91, pp. 167-172.

Tomizawa, T., Koshiba, S., Inoue, M., Kigawa, T., Yokoyama, S. & Initiative, R. S. G. P. (2008). pdb database 2YRO.

van den Brule, F., Califice, S. & Castronovo, V. (2004). Expression of galectins in cancer: A critical review. *Glycoconjugate Journal*, 19, pp. 537-542.

Vasta, G. R. (2009). Roles of galectins in infection. *Nature Reviews Microbiology* 7, pp. 424-438.

Wen, Y. F., Makagiansar, I. T., Fukushi, J., Liu, F. T., Fukuda, M. N. & Stallcup, W. B. (2006). Molecular basis of interaction between NG2 proteoglycan and galectin-3. *Journal of Cellular Biochemistry*, 98, pp. 115-127.

Yamamoto, H., Nishi, N., Shoji, H., Itoh, A., Lu, L. H., Hirashima, M. & Nakamura, T. (2008). Induction of cell adhesion by galectin-8 and its target molecules in Jurkat T-cells. *Journal of Biochemistry*, 143, pp. 311-324.

Yamaoka, A., Kuwabara, I., Frigeri, L. G. & Liu, F. T. (1995). A Human Lectin, Galectin-3 (Epsilon-Bp/Mac-2), Stimulates Superoxide Production by Neutrophils. *Journal of Immunology*, 154, pp. 3479-3487.

Yoshii, T., Fukumori, T., Honjo, Y., Inohara, H., Kim, H.-R. C. & Raz, A. (2002). Galectin-3 phosphorylation is required for its anti-apoptotic function and cell cycle arrest. *The Journal of Biological Chemistry*, 277, pp. 6852-6857.

Zhou, Q. & Cummings, R. D. (1993). L-14 lectin recognition of laminin and its promotion of in vitro cell adhesion. *Arch Biochem Biophys*, 300, pp. 6-17.

Zhuo, Y. & Bellis, S. L. (2011). Emerging Role of alpha 2,6-Sialic Acid as a Negative Regulator of Galectin Binding and Function. *The Journal of Biological Chemistry*, 286, pp. 5935-5941.

Zick, Y., Eisenstein, M., Goren, R. A., Hadari, Y. R., Levy, Y. & Ronen, D. (2004). Role of galectin-8 as a modulator of cell adhesion and cell growth. *Glycoconjugate Journal*, 19, pp. 517-526.

2

Nanostructural Chemically Bonded Ca-Aluminate Based Bioceramics

Leif Hermansson
Doxa AB
Sweden

1. Introduction

Biomaterials are based on a broad range of materials, such as organic polymers, metals and ceramics including both sintered and chemically bonded ceramics (silicates, aluminates, sulphates and phophates). The biomaterials can be made prior to use in the body in a conventional preparation of the material. The need for in situ in vivo formed implant materials makes the chemically bonded ceramics especially potential as biomaterials. These ceramics include room/body temperature formed biomaterials with excellent biocompatibility. Ca-aluminate as a biomaterial has been evaluated for over two decades with regard to general physical, mechanical and biocompatible properties. The Ca-aluminate based materials exhibit due to their unique curing/hardening characteristics and related microstructure a great potential within the biomaterial field. The presentation in this chapter aims at giving an overview of the use of Ca-aluminate (CA) as a biomaterial within odontology, orthopaedics and as a carrier material for drug delivery. The examination deals with aspects such as; the chemical composition selected, inert filler particles used, early properties during preparation and handling (working, setting, injection time, translucency, radio-opacity), and final long-term properties such as dimensional stability and mechanical properties (fracture toughness, compressive and flexural strength, hardness and Young´s modulus). One specific topic deals with the sealing of the Ca-aluminate biomaterials to tissue - a key in the understanding of the mechanisms of nanostructural integration.

2. Overview of properties of chemically bonded ceramics

The Ca-aluminate bioceramics belong to the chemically bonded ceramics, which are usually presented or known as inorganic cements (Mangabhai, 1990). Three different cement systems – Calcium phosphates (CP), Calcium aluminates (CA) and Calcium silicates (CS) are discussed in some details in this section. Ceramic biomaterials are often based on phosphate-containing solubable glasses, and various calcium phosphate salts (Hench, 1998). These salts can be made to cure *in vivo* and are attractive as replacements for the natural calcium phosphates of mineralised tissues. The Ca-phosphate products are gaining ground in orthopaedics as resorbable bone substitutes. Biocements are often based on various calcium phosphate salts – sometimes in combination with Ca-sulphates (Nilsson, 2003). These salts can be made to cure *in vivo* and are attractive as replacements for the natural mineralised tissues. However, these products have low compression strength values - in the interval 10-40 MPa - and are therefore questioned as load bearing implants.

Materials based on Ca-aluminate (CA) and Ca-silicate (CS) with chemistry similar to that of Ca-phosphates (CP) contribute to some additional features of interest with regard to dental and orthopaedic applications (Scriviner, 1988). The inherent difference in water uptake between the CA/CS-systems and CP gives benefits as:

- Higher mechanical strength.
- Possibility to add fillers, e.g. for improved radio opacity.
- Tuneable handling properties, e.g. rheology.

Materials based on Ca-aluminate and Ca-silicate thus contribute to some additional features of interest with regard to dental and orthopaedic applications. These features are related to the high amount of water involved in the curing process, the early and high mechanical strength obtained, and the biocompatibility profile including *in situ* reactions with phosphates ions of the body fluid. Compressive strength of cements based on Ca-silicate and Ca-aluminate is in the range 50-250 MPa depending on the water to cement ratio. The mechanical properties tested for the three Ca-based cement systems are compiled in Table 1 (Kraft, 2002, Loof et al 2003, Engqvist et al 2006).

Property profile after 7 days	Ca-aluminate based material	Ca-silicate based material	Ca-phosphate based material
Compressive strength (MPa)	100-200	100-150	< 100
Flexural strength (MPa)	30-60	30-40	20-30
Young´s modulus (GPa)	Approx. 15	Approx. 11	Approx. 3

Table 1. Mechanical property profile of the Ca-aluminate, Ca-silicate and Ca-phosphate systems

The water content involved in the hydration of the different chemically bonded ceramics are presented in Table 2.

System	Typical phase(s)	Oxide formula	Mol % H_2O	Weight-% in hydrated product
Ca-phosphate	Apatite	$10CaO\ 3P_2O_5\ H_2O$	7	Approx 5
Ca-aluminate	Katoite + gibbsite	$3CaO\ Al_2O_3\ 6\ H_2O$ + $Al_2O_3\ 2H_2O$	> 60	Approx 25
Ca-silicate	Tobermorite + amorphous phases	$5CaO\ 6SiO_2\ 5H_2O$ + Ca, Si)H_2O	> 30	Approx 20

Table 2. The three chemically bonded ceramic systems most used for biomaterials

2.1 Materials and basic function
2.1.1 Main chemistry

The injectability and handling features of the chemically bonded ceramics is mainly caused by the added water as the reacting phase with the powdered cements. This reaction is an acid-base reaction where water acts as a weak acid and the cement powder as a base. Several cement phases exist in the CaO - Al_2O_3 (CA) and in the CaO-SiO_2 (CS) systems, see Figure 1, but only a few are suitable as injectable materials. For one of the most attractive phases –

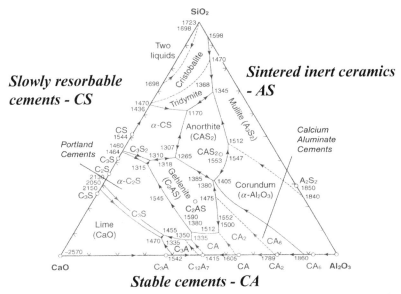

Fig. 1. The phase diagram of the high-strength bioceramic cements (Muan and Osbourne, 1965)

the mono Ca-aluminate, $CaOxAl_2O_3$ - the reaction at above 30 °C is described using cement chemical abbreviation as (Mangabhai, 1990);

$$3CA \quad + \quad 12H \quad \rightarrow \quad C_3AH_6 + 2AH_3 \tag{1}$$

Or

$$3\,(CaAl_2O_4) + 12\,H_2O \rightarrow Ca_3(Al(OH)_4)_2(OH)_4 + 4\,Al(OH)_3 \tag{2}$$

This example demonstrates 1) the phases obtained, C_3AH_6 (katoite) and AH_3 (gibbsite), and 2) the amount water consumed in the reaction. The technological importance of this is that all the water needed for paste like consistency or injectability can be consumed in the formation of solid phases yielding products with low porosity, one of the requirements for high strength. As can be seen in Eq. (1) three $CaAl_2O_4$ units consume twelve water molecules during hardening. This can be compared with the hardening of Calcium Phosphate Cement (CPC) where practically no extra water is consumed. In the case of CPC, the reaction liquid is only used as a vehicle for the reaction to take place. In both cases the hardening and formation of a solid body are driven by the precipitation of small hydrates, nano-size crystals of katoite and gibbsite for CA, and apatite or brushite for CPC. The amount required water is as high as 28% for CA. The water content of apatite, $Ca_5(PO_4)_3OH$, or $10CaOx3P_2O_5\ H_2O$ ($C_{10}P_3H$) is 2%. The CPC-system may contain some additional Ca-sulphate which can pick up some extra water. An additional 30 % Gypsum, $CaSO_4\ \frac{1}{2}\ H_2O$ to the CPC-system yields approximately a total water content of 6%.

The rate of the hydration is controlled by 1) the cement phase, 2) the particle size of the cement, 3) the hydration temperature and 4) processing agents including especially accelerators. Typical powder composition data for CA-cements as biomaterials are shown in

Table 3. Zirconia or high-density glasses are added for achievement of increased strength and increased radio-opacity. The glasses are used preferentially in dental applications where translucency is an additional desired feature. The high radiopacity of zirconia or other heavy-element containing phases means that the physician during the injection can follow the paste penetration in bone tissue without risking any possible leakage of the material into the surrounding tissue.

Compound	Formula	Function	Amount (wt-%)	Mean particle size
Ca-aluminate	$CaOxAl_2O_3$	Cement binder	50-70	< 5 μm
Zirconium dioxide	ZrO_2	Radiopacier	20-40	< 1 μm
μ-Silica	SiO_2	Expansion and viscosity controller	< 10	<50 nm

Table 3. Typical composition of an injectable biomaterial cement powder.

Typical processing agents are accelerators/retarders, dispersants, viscosity agents to control reaction rate, temperature and the cohesiveness, and in general the rheology. Examples are lithium chloride, polycarboxylate polymers and cellulose, as well as glass poly-alkeonates. For the CS-system Ca-chloride at high concentrations is normally used as an accelerator. For cements as injectable biomaterials, the reaction rate must be controlled with respect to working time, setting time, curing time and the maximum temperature during hydration. Typical data are presented in Table 4. The cement reactions are all exothermic and the temperature raise is controlled by the specific cement phase selected, and the hydration rate and the amount of material injected. For dental application the temperature raise is limited to a few °C above 37 °C. For orthopaedic applications where larger amounts (2-10 cm³) are used the temperature raise is more pronounced but lower than that of the conventional PMMA-based materials (Lewis, 2006).

System	Working time at 23 °C, min	Setting time at 37 °C, min	Max reaction temperature, °C
Ca-aluminate	Approx. 5	8-12	< 60, (for dental applications < 40)
Ca-silicate	Approx. 10	15-18	< 45
Ca-phosphate	5	10-12	< 40
PMMA	5-10	11	< 90

Table 4. Typical working and setting times and maximum reaction temperature of the systems discussed.

The on-going precipitation of hydrates and the reduction of the amount of liquid phase result in the formation of a material skeleton. This repeating reaction is fast at the beginning, resulting in a hardened product within 4-20 minutes depending on intended application. Strength corresponding to load carrying capacity is reached after approximately one hour.

However, the final strength and in the case of dental applications the translucency (Engqvist et al 2004), are reached after approximately a few days maturing.

It should be noted that for CBC materials, irrespective of chemistry (CA, CS or CPC), the final porosity cannot be zero. When the hydrates precipitate there is a contraction of approximately 10 %. The porosity is in the nanometer range and its exact amount is difficult to determine, but < 10 % of the filled space (the original liquid-phase volume) is estimated to be pores. However, this internal shrinkage related to hydration and precipitation yield no total shrinkage of the bulk material. In contrast, a limited expansion close to zero, may be induced (Kraft 2002). The porosity lowers the mechanical strength (although being in the nanometer range) but also enables liquids to diffuse into, or even through, the hardened CBC materials. This is an important feature when release of loaded drugs is a complementary aspect of the injectable biomaterial.

3. Ca-aluminate – general description and property profile

Ca-aluminates comprise double oxides of CaO and Al_2O_3. Several intermediate phases exist and these are - using the cement chemistry abbreviation system - C_3A, $C_{12}A_7$, CA, CA_2 and CA_6, where C=CaO and A=Al_2O_3. See Fig. 1 above. Table 5 presents typical property data.

Property	Typical value	Interval*
Compression strength, MPa	150	60-270
Young´s modulus, GPa	15	10-20
Thermal conductivity, W/mK	0.8	0.7-0.9
Thermal expansion, ppm/K	9.5	9-10
Flexural strength, MPa	50	20-80
Fracture toughness, MPam$^{1/2}$	0.5	0.3-0.8
Corrosion resistance, water jet impinging, reduction in mm	< 0.01	-
Radio-opacity, mm	1.5	1.4-2.5
Process temperature, ^{o}C	> 30	30-70
Working time, min	3	< 4
Setting time, min	5	4-7
Curing time, min	20	10-60
Porosity after final hydration, %	15	5-60

The interval is primarily related to the c/w ratio used, and the highest values are achieved with c/w ratio close to that of complete hydration with no excess of water

Table 5. Mean property data of dental Ca-aluminate based materials (Kraft 2002, Lööf 2008, Lööf et al 2004,2005, Hermansson et al 2008)

Due to reduced porosity based on the huge water uptake ability, the Ca-aluminate material exhibit the highest strength among the chemically bonded ceramics. The inherent flexural strength is above 100 MP based on measurement of the fracture toughness, which is about 0.7 - 0.8 MPam$^{1/2}$. The actual flexural strength is controlled by external defects introduced during handling and injection of the material. The thermal and electrical properties of Ca-aluminate based materials are close to those of hard tissue, the reason being that Ca-aluminate hydrates chemically belong to the same group as Ca-phosphates, the hard tissue of bone. Another important property related to Ca-aluminate materials is the possibility to control the dimensional change during hardening. In contrast to the shrinkage behaviour of many polymer-based biomaterials, the Ca-aluminates exhibit a small expansion, 0.1-0.3 linear-% (Kraft 2002).

3.1 Biocompatibility including bioactivity

Definitions used

The terms biocompatibility and bioactivity are used in different ways by different categories of scientists. Below are presented the definitions used in this paper, mainly agreeing with the definitions discussed in (Williams, 1987) Biocompatibility: "The ability of a material to perform with an appropriate host response in a specific application".
Bioactivity (bioactive material): "A material which has been designed to induce specific biological activity". Another definition according to (Cao and Hench,1996) "A bioactive material is one that elicits a specific response at the interface of the material which results in the formation of a bond between the tissues and the material".
Thus a material cannot in itself be classified as biocompatible without being related to the specific application, for which it is intended. Bioactivity from a materials viewpoint is frequently divided into *in vitro* and *in vivo* bioactivity. The *in vitro* bioactivity of a material is however only an indication that it might be bioactive in a specific *in vivo* application. Another aspect of bioactivity is that this term can be adequate only when the biomaterial is in contact with a cellular tissue. However, often a material is claimed to be bioactive if it also reacts with body liquids forming an apatite-phase. *In vitro* bioactivity is tested in phosphate buffer systems similar to that of saliva or body liquid, and apatite formation is the claimed sign of bioactivity. A further aspect of bioactivity and also biocompatibility deals with the different curing times and temperatures at which the observation (testing) is performed. This is important to issues such as initial pH-development, cohesiveness and initial strength. Finally the biocompatibility and bioactivity can only be confirmed in clinical situations, with the actual implant/biomaterial in the designed amount or content and shape. This is especially important for injectable biomaterials which are formed (hydrated) and cured *in vivo*, and for implants where movements, even micro-movement, can influence the outcome.

Standards and methods

Relating to the definition aspects above, the acceptance of a biomaterial is a crucial issue, and to some extent the question has been solved by relating to the following toxicological endpoints indicating biocompatibility as referred in the harmonized standard ISO 10993:2003, which comprises the following sections:
Cytotoxicity (ISO10993-5), Sensitization (ISO10993-10), Irritation/Intracutaneous reactivity (ISO10993-10), Systemic toxicity (ISO10993-11), Sub-acute, sub-chronic and chronic toxicity (ISO10993-11), Genotoxicity (ISO10993-3), Implantation (ISO10993-6), Carcinogenicity (ISO10993-3) and Hemocompatibility (ISO10993-4).
This will be the main guideline when presenting the status of the biocompatibility of the CASPH-system, but was complemented by corrosion testing, elementary analysis, pH-change and additional cytotoxicity testing.
The corrosion resistance test – using a water jet impinging technique - was conducted according to EN 29917:1994/ISO 9917:1991,where removal of material is expressed as a height reduction using 0.1 M lactic acid as solution, pH 2.7 . The duration time of the test is

24 h. The test starts after 24 h hydration. The test probe accuracy was 0.01 mm. Values below 0.05 mm per 24 h solution impinging are judged as acid resistant.

Determination of Ca and Al in the solution during the hydration process of the Ca-aluminate based material was performed using atomic absorption spectrometry (Liu et al 2002). Standard solutions of different concentrations of Ca and Al were prepared according to the manual. Samples were prepared with a size of 10mm x 2mm height using a wet-press method, corresponding to a surface area of 224 mm². The test pieces were placed in plastic bottles in inorganic saliva solution of pH 7. The amount of liquid was 10 ml in each bottle. The temperature selected was 37 ºC. The inorganic saliva solution contained calcium chloride, magnesium chloride, sodium chloride, a phosphate buffer, hydro-carbonate and citric acid. The Ca-content in the saliva solution corresponded to 68 ppm. 1 ml solution was removed at 1, 7 and 28 days for analyses, and saliva was exchanged at 1, 7 and 28 days after every measurement. For the 28 days test additional samples were also taken 1 h after new solution was added.

Measurement of pH development during hydration of the material was conducted using a standard pH-meter. Samples were prepared according to the procedure for atomic absorption described above. The pH-testing was conducted in two separate ways. First the samples were immersed in saliva solution (pH =7) at 37 ºC, and pH was measured continually over the whole experiment period (Test 1). 1 ml solution was removed at 1, 7, 14 and 28 days for pH measurement. The second type of measurement comprised immersion in 10 ml saliva solution at 37 ºC, where the saliva was exchanged at 1, 7, 14 and 28 days. pH was measured at the time of observation and also after one hour in new saliva (data within brackets), Test 2.

Specimens at different setting stages were subjected to cytotoxicity testing by using primary cultures of human oral fibroblasts. A tissue culture insert retaining tested materials was assembled into a 12-well plate above the fibroblast monolayers. The cytotoxicity was determined by MTT reduction assay after various curing times. Specimens were set and hydrated at 37 ºC for different periods of time, i.e. 0, 5, 30, 60 min, 24 h and 1 week and were then placed on tissue culture netwell for a cytotoxicity test. Both acute (1 and 24 h) and long term (1 week) in vitro toxicity tests were conducted with MTT assay.

3.1.1 Biocompatibility including bioactivity of Ca-aluminate bioceramics

Biocompatibility evaluation

Summarized below are the results from several biocompatibility and bioactivity studies (Engqvist 2004, 2005, Faris 2006) where Ca-aluminate is used as a biomaterial in orthopaedic and dental applications. In vitro bioactivity studies show apatite formation on the surface of the Ca-aluminate materials exposed to phosphate buffer solution, an example shown in Figure 2 below.

Corrosion resistance

No height reduction at all was observed for two tested Ca-aluminate materials. Thus, according to the acid corrosion test, Ca-aluminate materials are judged as stable materials. The total absence of material loss, measured as height reduction in the acid corrosion test, is related to the general basic nature of the material, with possibility of neutralization of the acid in the contact zone – especially in the earlier stage of the hydration process.

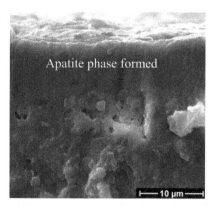

Fig. 2. Cross section of the apatite-containing surface layer formed, SEM (Engqvist et al, 2004)

Ion release measured by atomic absorption spectrometry

The results of Ca and Al determination in the solution during the hydration process of the Ca-aluminate based material are presented in Table 6.

Ion tested	1h, ppm	24 hrs, ppm	7 days, ppm	28 days, ppm
Ca	66	64	44	50 (70)
Al	11.4	9.3	9.6	8.6 (1.2)

Table 6. Ca and Al dissolution during hardening of the Ca-aluminate material, (The 1h testing at 28 days within brackets)

The release of metal ions in water was below $5x10^{-2}$ ppm/(mm^2 material) for aluminium and below $30x10^{-2}$ ppm/(mm^2 material) for calcium, whereas somewhat higher aluminium content was measured in artificial saliva. The ion concentrations detected are generally not time-dependent during hydration. After the initial hydration time the ion concentration (molar) is determined by the solubility product of the phases formed (katoite = $5x10^{-26}$ and gibbsite = $3x10^{-24}$). Since the concentration of Ca in saliva is higher than what is obtained in the non-physiological aqueous solution (distilled water), it can be assumed that the filling material releases very limited amounts of Ca or Al once the material has hardened. The presence of Ca in saliva will decrease the solubility tendency of the calcium-aluminate-hydrate phases.

Based on a search in the literature, the FAO/WHO Joint Expert Committee on Food Additives (JECFA) has provided a provisional figure for tolerable weekly aluminium intake of 7 mg/kg body weight. This corresponds to 1 mg/kg /day. The daily intake of aluminium via digestion/food is approximately 5 mg per day. For calcium the NIH Consensus Development Conference on Optimal Calcium Intake recommended an intake in the range 800 mg/day for young children to 1000 – 1500 mg/day for adults depending on gender and age. For many people there is a need to supply additional calcium in order to stay healthy. The ion concentrations measured and the amounts of Ca and Al released are far below the concentrations of the elements produced from food intake and should therefore not pose any safety concerns at all.

Change in pH during hydration

The results of the measurement of pH development during hydration of the Ca-aluminate based are shown in Table 7. The initial *in vitro*-pH is 10.5 in saliva. After 1 week, pH after 1 h dissolution time in saliva is approx. 8.

Test No.	At start	1h	24 hrs	7 days	14 days	28 days
1	10.5	10.3	10.7	10.3	9.9	9.8
2	10.5	10.2	10.2 (7.7)	9.9 (7.8)	9.5 (8.1)	9.2 (8.1)

Table 7. Change of pH during initial hydration of Ca-aluminate based materials

The pH is high during the early stage of the hydration, but decreases with time and approaches neutrality. The reason for the high pH in the beginning is the general basic character of the material and the formation of OH- during the hydration process. In the clinical situation saliva is produced in a dynamic way, creating an environment capable of buffering surrounding solution to neutrality. In the clinical studies performed so far no adverse reactions have been reported from a possible elevated pH during the early part of the hydration.

When Ca-aluminate material is combined with glass ionomer system the pH-system becomes initially acidic. However after 10 min the pH is above neutral, but will not exceed pH 9 (Jefferies et al 2009).

Cytotoxicity testing

The *in vitro* MTT reduction test of the experimental Ca-aluminate dental filling material in human oral fibroblast culture showed no obvious cytotoxicity. The average level of MTT reduction of the experimental dental filling material was close to 100% of the control values. The maximal variation (SD) was less than 30%. Different curing times of the test material did not seem to affect the cytotoxicity test results although one week curing produced the most stable testing results both in the short and long term tests. After a week the material can be considered as fully cured, i.e. stable.

Morphological changes were not observed in any of the test groups at different MTT reduction testing points. As shown in Figure 3, the cell culture was typically fibroblastic with a slender and elongated form in both the control group and the group exposed to the examined material. In the exposed picture B even some precipitated hydrates are seen.

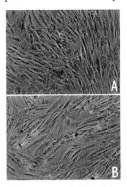

Fig. 3. Morphological observation of human oral fibroblasts on an experimental Ca-aluminate based material. A: Normal control. B: After exposure to the experimental filling material for one week (Liu et al 2002).

The standard for cellular biocompatibility in *in vitro* testing has been stated in the International Organization for Standardization (ISO) standards documents. The standard allows for the contact testing of solid dental materials for cytotoxicity with cell lines. Due to several disadvantages of direct contact testing, indirect testing methods have been developed and compared to the direct testing assays (Tang et al 2002). Introduction of a standard cell culture device, i.e. cell culture insert or transwell, provided an opportunity for such cytotoxicity screening of dental materials with indirect contact between material specimens and cell culture monolayer. It is believed that such a testing system more closely mimics the *in vivo* exposure pattern by providing the test of the material in both its solid and dissolved phases at the same time. It has been shown that this testing system has produced the most stable results as compared to other testing systems, such as direct contact test. In a complementary cytotoxicity test using the pulp derived cell response, the experimental CA-material showed no sign of toxicity (Schmalz 2002).

Harmonized standard ISO 10993:2003

Further cytotoxicity and other biocompatibility aspects are summarised according to the outline in the harmonized standard ISO 10993:2003. An experimental orthopaedic Ca-aluminate-based material was the test material. This material is judged as mildly cytotoxic during the initial curing, and as non-cytotoxic as cured material. See Table below.

Type of test	Method	Cytotoxicity (scale 0-4 or 100-0%
During curing, undiluted	ISO 10993-5, § 8.2	2 (mild)
During curing, diluted	ISO 10993-5, § 8.2	0-1 (none-slight)
During curing	XTT-test	60 % (slight)
During curing, diluted	XTT-test	> 70 % (none)
Cured, undiluted	ISO 10993-5, § 8.2	0 (none)
Cured, diluted → diluted	XTT-test	> 70% (none)

Table 8. Cytotoxicity testing of an orthopaedic Ca-aluminate based material

A sensitization test (ISO 10993-10), Guinea Pig Maximization Test was performed with the orthopaedic Ca-aluminate material during curing. No sensitizing potential was obtained. Additional irritation and delayed hypersensitivity testing according to ISO 10993-10:2002 was conducted with both polar and non-polar extract from cured material, and the results showed no discrepancies after intracutaneous injections in the rabbit compared to the blank injections. The acute systemic toxicity study according to ISO 10993-11 was performed with both polar and non-polar extracts from cured Ca-aluminate material (Xeraspine), and the results showed no signs of acute systemic toxicity. Sub-acute, sub-chronic and/or chronic toxicity studies according to ISO 10993-11 were not conducted explicitly, since data from the two implantation studies in rabbit (see below) were judged to support that no long term toxicity is expressed. From the implantation studies histopathological organ and tissue data is available, and no adverse effects were reported.

Additionally, in an *in vivo* genotoxicity assay, the mice micronucleus test of bone marrow was used. The extract (The experimental Ca-aluminate material during curing and cured material) was administered intraperitoneally twice. The results showed no clastogenic effect. Three *in vivo* implantation studies based on ISO 10993-6 have been performed. Two studies in rabbit (femur) and one in sheep (vertebrae). *In vivo* implantation studies are judged as the most relevant studies for documentation of safety of a product. In the rabbit implantation studies Ca-aluminate material was compared to the PMMA-material CMW 1, and in the sheep study Ca-aluminate material was compared to the PMMA-material Vertebroplastic, and to the Bis-GMA material Cortoss™. The results are summarized in Table 9.

Implantation studies	Species	Reference material	Result
6-week femur	Rabbit	CMW-1	Minimal inflammation, very few inflammatory cells were present in bone, bone marrow and adipose tissues.
6 (12) -month femur	Rabbit	CMW-1	No Al- accumulation
12-week vertebrae	Sheep	Vertebroplastic Cortoss	No inflammation, no Al-accumulation

Table 9. Implantation studies in femur rabbit, and in vertebrae sheep, details in (Hermansson et al 2008)

The 6 months femur study in rabbits included a 12 months subgroup. The amount of aluminium in blood and selected organs was analysed. The main target organs of the animals (kidney, lung, liver) were histopathologically investigated. Granulomatous inflammation in the cavity, pigmented macrophages and new bone formation were the treatment-related observations at 6- and12-months examination. No difference between Ca-aluminate material and PMMA was detected. There were no signs of aluminium accumulation in the analysed tissues.

In the 12-week study, the histopathology of vertebrae obtained one week after surgery showed the most severe inflammatory reaction to the surgery in the *sham* operated vertebrae. The bone marrow in the vertebrae filled with Ca-aluminate was not reported to be infiltrated by any inflammatory cells. In vertebrae obtained 12 weeks after surgery no inflammatory reactions were reported, and no obvious differences were observed in the pathological reactions to the surgery (sham) or the filler materials. Overview of the histological contact zone to the Ca-aluminate based material is shown in Figure 4.

The analysis of serum samples showed low concentrations of aluminum in comparison to what is normal in humans. Since the concentration of aluminum did not increase after surgery and in some instances was lower after surgery than in the 0-samples, one may regard these concentrations as within the normal variation.

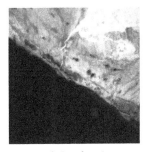

Fig. 4. Histology image of an experimental Ca-aluminate material (black) in close contact with sheep vertebral bone.

Repeated haemocompatibility studies have been performed to evaluate possible reactions in whole human blood as a result of contact with Ca-aluminate materials (Axen et al 2004). Test items were an experimental Ca-aluminate based material and Xeraspine, Vertebroplastic and Norian (Calcium Phosphate Cement, Synthes Inc). A Chandler loop model was used in which circulating human blood was in contact with the test materials for up to 4 hours. For comparison, loops free from test materials were used. Platelet count (PLT), thrombin-antithrombin (TAT) complex, complement factors C3a and C5b-9 (TCC), and TNF-α were assayed. The degree of haemolysis was assessed by the Drabkin method. Norian (a calcium phosphate based material) invariably induced extensive clotting already after 60 minutes, verified macroscopically and also by significantly reduced PLT in comparison to the Control loops, whereas there was no significant reduction in PLT in the loops with Ca-alumiante material or Vertebroplastic, respectively, neither at 60 nor at 240 minutes. The Ca-aluminate material did not induce haemolysis to a greater extent than any of the other materials tested. TCC was activated to a certain degree by the biomaterial, comparable to what is commonly observed for artificial materials. TNF-α generation, indicative of activation of white blood cells, was not enhanced by either Vertobroplastic or the Ca-aluminate material.

Based on all above mentioned data and generated toxicity data, it is considered that there is no reason to expect that the Ca-aluminate biomaterials when used in accordance with the intended clinical use will create any adverse effects. The Ca-aluminate based materials fulfill the requirements of the harmonized standard ISO 10993:2003.

3.1.1.1 Complementary reactions of Ca-aluminate in presence of body liquid.

Complementary reactions occur when the Ca-aluminate is in contact with tissue containing body liquid. Several mechanisms have been identified, which control how the Ca-aluminate material is integrated onto tissue. These mechanisms affect the integration differently depending on what type of tissue the biomaterial is in contact with, and in what state (un-hydrated or hydrated) the CA is introduced. These mechanisms are summarized as follows and described in more details elsewhere (Hermansson, 2009);

Mechanism 1:   Main reaction, the hydration step of CAC (Eq. 1 above)
Mechanism 2:   Apatite formation in presence of phosphate ions in the biomaterial
Mechanism 3:   Apatite formation in the contact zone in presence of body liquid
Mechanism 4:   Transformation of hydrated Ca-aluminate into apatite and gibbsite
Mechanism 5:   Biological induced integration and ingrowth, i.e. bone formation at the contact zone
Mechanism 6:   Point-welding due to mass increase when in contact with body liquid.

When phosphate ions or water soluble phosphate compounds are present in the biomaterial (powder or liquid) an apatite formation occurs according to the reaction

$$5Ca^{2+} + 3PO_4^{3-} + OH^- \rightarrow Ca_5(PO_4)_3OH$$

This complementary reaction to the main reaction occurs due to the presence of Ca-ions and a basic (OH^-) environment created by the Ca-aluminate material. The solubility product of apatite is very small [$Ks = 10^{-58}$], so apatite is easily precipitated. Body liquid contains among others the following ions HPO_4^{2-} and $H_2PO_4^-$. In contact with the Ca-aluminate system and water during setting and hydration, the presence of Ca-ions and hydroxyl ions, the hydrogen phosphate ions are neutralised according to

$$HPO_4^{2-} + H_2PO_4^- + OH^- \rightarrow PO_4^{2-} + H_2O,$$

whereafter the apatite-formation reaction occurs

$$5Ca^{2+} + 3PO_4^{3-} + OH^- \rightarrow Ca_5(PO_4)_3OH$$

This reaction occurs upon the biomaterial surface/periphery towards tissue. The apatite is precipitated as nano-size crystals (Hermansson et al, 2006). See figure 5.

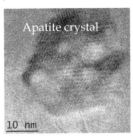

Fig. 5. Nano-size apatite formation in the the contact zone to hard tissue

Katoite is formed as a main phase, and is kept as katoite in the bulk material according to the mechanism 1 above. However, in long-time contact with body liquid containing phosphate ions the katoite is transformed at the interface tobody tissue into the at neutral pH even more stable apatite and gibbsite phases according to

$$Ca_3 \cdot (Al(OH)_4)_2 \cdot (OH)_4 + 2\,Ca^{2+} + HPO_4^{2-} + 2\,H_2PO_4^- \rightarrow$$

$$Ca_5 \cdot (PO_4)_3 \cdot (OH) + 2\,Al(OH)_3 + 5\,H_2O$$

When apatite is formed at the interface according to any of the reaction mechanisms 2-4 above, at the periphery of the bulk biomaterial, the biological integration may start. Bone ingrowth towards the apatite allows the new bone structure to come in integrated contact with the biomaterial. This is an established fact for apatite interfaces. For the CA-system the ingrowth is discussed below, 4.4. The transition from tissue to the biomaterial is smooth and intricate.

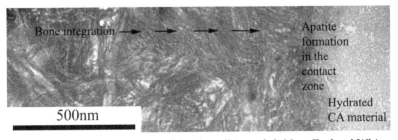

Fig. 6. Integration of CA in tissue – a model using albino adult New Zealand White rabbits (Hermansson et al, 2008).

The actual contact zone developed depends on a combination of the above discussed mechanisms and the tissue. The latter varies from a cellular-free high content apatite tissue in the case of a dental enamel, via dentine to a bone structure with cellular and body liquid contact. Also the material can be in contact with other implant materials as dental crowns, dental screws or coatings on implants. In the tables 10 and 11 are summarized in which applications and specific tissues the demonstrated mechanisms are predominant.

Tissue	Mech 1	Mech 2	Mech 3	Mech 4	Mech 5	Mech 6
Enamel	x	x				
Dentine	x	x	X	x	(x)	
Bone	x	x	X	x	x	x

Table 10. Type of tissue and possible mechanisms.

Application	Mech 1	Mech 2	Mech 3	Mech 4	Mech 5	Mech 6
Cementation a. towards tissue	x	x	X	x	x	
b. towards implant	x	(x)				
Dental fillings a. towards enamel	x	x				
b. towards dentine	x	x	X	x	(x)	
Endo fillings a. orthograde	x	x				
b. retrograde incl bone	x	x	X	x	x	
Coatings and augmentation towards implant	x	(x)				
gap filling	x	x	X	x	x	x

Table 11. Applications and possible mechanisms.

Nanostructure and nano-porosity used in certain applications

The nanostructure including nanoporosity developed in the Ca-aluminate biomaterial system when near complete hydration occurs, yields some unique properties related to how bacteriostatic and antibacterial properties may develop in the biomaterial. The nano-porosity can also be used to control release of drugs incorporated the biomaterial. The background to this is that even if the total porosity is low, all porosity is open, thus allowing transport of molecules in the nanoporosity channels. The nanostructure used in thin film coatings will also be touched upon.

Antibacterial aspects

The surprising finding in studies recently performed (Doxa patent application 2010) shows that the bacteriostatic and antibacterial properties of the Ca-aluminate biomaterial are not primarily related to pH or specific ions and ion concentration or reducing agents, but to the hydration procedure and the microstructure obtained. This also to some extent is an answer why highly biocompatible and even bioactive biomaterials can combine apparently contradictory features such as biocompatibility, bioactivity and apatite formation and environmental friendliness with bacteriostatic and antibacterial properties.

The bacteriostatic and antibacterial properties are primarily related to the development of the nanostructure and the nano-size porosity during hydration of the Ca-aluminate system. The initial low pH (< 8) of the system in the case of the presence of a polycarboxylic acid for cross-linking, is such not a hindrance for the antibacterial properties. The requirements of the microstructure of Ca-aluminate and/or Ca-silicate based biomaterials to achieve antibacterial properties are related to the general nanostructure obtained; A nanoparticle/crystal size of hydrates in the interval 15-40 nm, a nanoporosity size of 1-4 nm and the number of pores per square micrometer of at least 500, preferably > 1000.

The above mentioned requirements will guarantee that the nanostructure will be free of large pores meaning no escape of bacteria within the original liquid, paste or dental void, during the hydration. The nanocrystals will participate on all walls, within the liquid, and on all inert particles and on bacteria within the original volume. The formation of nanocrystals will continue to all the void is filled. The bacteria will be totally encapsulated and will be chemically dissolved. Also the number of nanopores will be extremely which will have the possibility of catching and fasten bacteria to the hydrate surface – an analogue to how certain peptides may function as antibacterial material due to a structure with nanosize hole within the structure.

4. Materials and biomaterials application

Alternative dental materials and implant materials based on bioceramics are found within all the classical ceramic families: traditional ceramics, special ceramics, glasses, glass-ceramics, coatings and chemically bonded ceramics (CBC) (Ravaglioli and Krajewski, 1992). The CBC-group, also known as inorganic cements, is based on materials in the system CaO-Al_2O_3-P_2O_5-SiO_2, where phosphates, aluminates, and silicates are found. Depending on in vivo chemical and biological stability, the CBC biomaterials can be divided into three groups: stable, slowly resorbable and resorbable. The choice for dental and stable materials is the Ca-aluminate based materials (Hermansson et al 2008). Slowly resorbable materials are found within Ca-silicates and Ca-phosphates, and fast resorbing materials among Ca-sulphates and some Ca-phosphates. The stable biomaterials are suitable for dental applications, long-term load-bearing implants, and osteoporosis-related applications. For trauma and treatment of younger patients, the preferred biomaterial is the slowly resorbable materials, which can be replaced by new bone tissue (Nilsson, 2002). In this section are summarised some of the possible new applications using the strong chemically bonded ceramics based on Ca-aluminate. The presentation is devided in three application areas; dental, orthopaedics and drug delivery.

The following product areas have been identified based on experimental material data, pre-clinical studies, pilot studies and on-going clinical studies (Jefferies et al, 2009). The application areas are; Dental cement, endodontic products (orthograde and retrograde), sealants, restoratives, and pastes for augmentation and dental implant coatings. For low-viscosity and early hardening of the CA, a complementary glass ionomer can preferrably be used. Clinical use of the materials is foreseen within the next coming years. The use of CA within odontology is based on the following features; early/rapid anchoring, high strength, long-term stability, no shrinkage, combined bonding and bulk material, biocompatibility and in situ apatite formation ability (nanocrystals formed in the contact zone between material and tissue).

4.1 Dental applications

The existing dental materials are mainly based on amalgam, resin composites or glass ionomers. Amalgam, originating from the Tang dynasty in China, was introduced in the early 19th century as the first commercial dental material. It is anchored in the tooth cavity by undercuts in the bottom of the cavity to provide mechanical retention of the metal. Although it has excellent mechanical characteristics it is falling out of favor in most dental markets because of health and environmental concerns. One exception is the US in which amalgam still has a redlatively large market share.

The second generation material is the resin composites, first introduced in the late 1950s. These are attached to the tooth using powerful bonding agents that glue them to the tooth structure. After technical problems over several decades, these materials today have developed to a level where they work quite well and provide excellent aesthetic results. Despite the improvements, resin composites have some drawbacks related to shrinkage, extra bonding, irritant components, a risk of post-operative sensitivity, and technique sensitivity in that they require dry field treatment in the inherently moist oral cavity. The key problem, due to shrinkage or possible degradation of the material and the bonding, is the margin between the filler material and tooth, which often fails over time leading to invasion of bacteria and secondary caries. Secondary caries is a leading cause of restorative failure and one of the biggest challenges in dentistry today. As a significant number of dental restorations today are replacement of old, failed tooth fillings, it is clear that tackling this problem is a major market need (Mjör, 2000). Secondary caries occurs not only after filling procedures but also following other restorative procedures such as the cementation of crowns and bridges.

Glass ionomers were first introduced in 1972 and today are an established category for certain restorations and cementations. Their main weakness is the relatively low strength and low resistance to abrasion and wear. Various developments have tried to address this, and in the early 1990s resin-modified ionomers were introduced. They have significantly higher flexural and tensile strength and lower modulus of elasticity and are therefore more fracture-resistant. However, in addition to the problems of resin composites highlighted above, wear resistance and strength properties are still inferior to those of the resin composites.

The nature of the mechanisms utilized by Ca-aluminate materials (especially Mechanism 1 above) when integrating and adhering to tooth tissue and other materials makes these materials compatible with a range of other dental materials, including resin composite, metal, porcelain, zirconia, glass ionomers and gutta-percha. This expands the range of indications for Ca-aluminate based products from not only those involving tooth tissue, e.g. cavity restorations, but also to a range of other indications that involve both tooth tissue and other dental materials. Examples include dental cementation, base and liner and core build-up and endodontic sealer /filler materials, which involve contact with materials such as porcelain, oxides and polymers and metals, and coatings on dental implants such as titanium or zirconia-based materials.

The use of the Ca-alumiante materials may be a first step towards a paradigm shift for dental applications. The features are summarized below.

Nanostructural integration
- Reduced risk of secondary caries and restoration failure
- Excellent biocompatibility

- Reduced/no post-operative sensitivity, environmentally friendly and no allergy
- Optimized consistency and setting properties for the intended application
- Reliable clinical handling performance
- Hydrophilic acid-base system with no extra bonding or no dry field precaution
- Fast and easy clinical handling, no pre-treatment of tooth necessary
- Excellent retentive and mechanical properties, no degradation of properties over time
- Durable tooth-material restoration interface (longevity).

Below are presented in more details the state–of-the-art for different possible applications of Ca-aluminate based materials.

4.1.1 Dental cement

Long-term success after cementation of indirect restorations depends on retention as well as maintenance of the integrity of the marginal seal. Sealing properties of great importance deal with microleakage resistance, the retention developed between the dental cement and the environment, compressive strength and acid resistance. Data presented below support the Ca-aluminate-system as highly relevant for dental cement materials. Integration with tooth tissue is a powerful feature and the foundation of the Ca-aluminate technology platform. Secondary caries occurs not only after filling procedures but also after other restorative procedures such as the cementation of crowns and bridges. The consequence of the difference in the mechanism of action between Ca-aluminate products and conventional products is illustrated by the study presented in details in (Pameijer et al , 2008, 2009), illustrated in Figure 2 below. It shows that the micro leakage, measured by dye penetration after thermo cycling, of a leading dental cement (Ketac Cem®, 3M) was significantly higher, both before and after thermo cycling compared to Ceramir C&B, a Ca-aluminate based product recently approved by FDA. This has also recently been verified using techniques for studying actual bacterial leakage. The above described nanostructural precipitation upon tissue walls, biomaterials and within the original Ca-aluminate paste is the main reason for this, in addition to a high acid corrosion resistance.

General properties of the CAPH-system used as dental cement have been presented (Pameijer et al, 2008), see Fig. 7 and Table 12 below. General features of all the dental cement classes available are presented as a summary in Table 13 below.

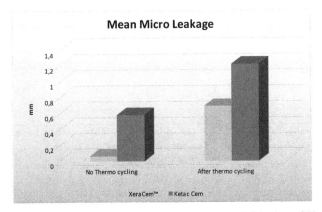

Fig. 7. Micro leakage leakage of a Ca-aluminate based material (blue) and Ketac Cem (red).

Material	Net setting time, min	Film thickness, μm	Compressive strength, MPa	Crown retention, Kg/Force
Ca-aluminate based	4.8	15	196 (at 30days)	38.6
Ketac Cem	-	19	-	26.6
RelyX-Unicem	-	-	157	39.4

Table 12. Selected properties, Test methods according to SO 9917-1

Material aspects	Glass Ionomer (GI)	Resin-modified GI	Resin (bonded)	Self-adhesive Resin	Zn-phosphate cement	Ceramir C&B
1. Type of material	Polymer	Monomer	Monomer	Monomer	Inorganic material	Ceramic-polymer
2. Hardening mechanism	Cross-linking	Poly-merisation	Poly-merisation	Poly-merisation	Acid-base reaction	Acid-base + cross-linking
3. pH	Acidic	Acidic	Acidic /neutral	Acidic /neutral	Acidic	Acidic /basic
4. Geo-metrical stability	Non-shrinking	Non-shrinking	Shrinks	Shrinks	Non-shrinking	Non-shrinking
5. Stability over time	Degrades	Degrades	Degrades	Degrades	Degrades	Stable
6. Extra treatment	-	-	Etching and bonding	-	-	-
7. Hydro-philic / phobic	Hydro-philic	Hydrophilic	Hydrophobic	Initially hydrophilic, Hydrophobic	Hydrophilic	Hydro-philic
8. Integration mechanism	Micro-mechanical retention, Chemical bonding	Micro-mechanical retention / Chemical bonding / Adhesion	Adhesion / Micro-mechanical retention	Adhesion / Micro-mechanical retention	Micro-mechanical retention	Nano-structural integration
9. General behaviour	Irritant	Allergenic	Allergenic	Allergenic	Non-allergenic	Non-allergenic
10. Biocom-patibility	Good	OK	OK	OK	Good	Good
11. Bioactivty	No	No	No	No	No	Bioactive
12. Sealing quality	OK	OK	Good but operation sensitive	OK	Acceptable	Excellent

Table 13. Overview of dental luting cements (Hermansson et al, 2010)

Material aspects 4-12 in Table 13 are also relevant for all other Ca-aluminate based dental applications.

A clinical 2-year study comprising 35 cemented crowns was conducted at Kornberg School of Dentistry, Temple University, and follow-up data and feedback from participating dentists were excellent with no failures at all reported (Jefferies et al 2009).

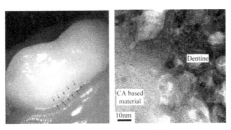

Fig. 8. Cemented ceramic crown (left), and HRTEM of the nanostructure of Ca-aluminate hydrates, hydrates formed in the interval 10-30 nm(right), bar = 10 nm (Hermansson et al, 2010)

4.1.2 Endodontics

In a review of the biocompatibility of dental materials used in contemporary endodontic therapy (Haumann and Love, 2003) amalgam was compared with gutta-percha, zinc oxide-eugenol (ZOE), polymers, glass ionomer cements (GICs), composite resins and mineral trioxide aggregate (MTA). A review (Niederman, 2003) of clinical trials of *in vivo* retrograde obturation materials summarized the findings. GIC's appeared to have the same clinical success as amalgam, and orthograde filling with gutta-percha and sealer was more effective than amalgam retrograde filling. Retrograde fillings with composite and Gluma, EBA cement or gold leaf were more effective than amalgam retrograde fillings. However, none of the clinical trials reviewed in included MTA. In a 12 week microleakage study, the MTA performance was questioned compared to that of both amalgam and a composite (Alamo et al, 1999).

The Ca-aluminate-based material discussed in this paper belongs to the same material group as MTA, the chemically bonded ceramics. MTA is a calcium silicate (CS) based cement having bismuth oxide as filler material for improved radio-opacity, whereas the Ca-aluminate material consists of Ca-aluminate phases CA and CA_2 with zirconia as filler material. MTA is claimed to prevent microleakage, to be biocompatible, to regenerate original tissues when placed in contact with the dental pulp or periradicular tissues, and to be antibacterial. The product profile of MTA describes the material as a water-based product, which makes moisture contamination a non-issue (Dentsply 2003). The CA-cement materials are more acid resistant than the CS-based materials, and in general show higher mechanical strength than the CS materials. A two-year and a five-year retrospective clinical study of Ca-aluminate based material have been conducted (Pameijer et al, 2004, Kraft et al, 2009). The study involved patients with diagnosis of either chronic per apical osteitis, chronic per apical destruction, or trauma. Surgery microscope was used in all cases. For orthograde therapy the material was mixed with solvent into appropriate consistency and put into a syringe, injected and condensed with coarse gutta-percha points. Machine burs were employed for root canal resection. For the retrograde root fillings, the conventional surgery procedure was performed. The apex was detected with surgery microscope and rinsed and prepared with an ultrasonic device. Crushed water-filled CA-tablets were then inserted and condensed with dental instruments. The patients' teeth were examined with X-ray, and three questions regarding subjective symptoms were put to patients: 1. Have you had any persistent symptoms? 2. Do you know which tooth was treated? 3. Can you feel any symptoms at the tooth apex?

In 13 of the 17 treated patients the diagnosis was chronic perapical osteitis (c p o). These were treated with retrograde root filling (rf) therapy. Three patients suffered from trauma or chronic perapical destruction, and these patients were treated with orthograde therapy. Out of 17 patients (22 teeth) treated, 16 patients (21 teeth) were examined with follow-up x-ray after treatment and also after two years or more. The additional patient was asked about symptoms. The results of both the clinical examination and the subjective symptoms were graded into different groups related to the success of the therapy. The results of the 2-year and the 5-year study are shown in Table 14.

	1 Complete healing		2 Incomplete healing		3 Uncertain		4 Failure	
	2-year	5-year	2-year	5-year	2 year	5-year	2-year	5-year
Nos. of teeth	18	14	3	2	0	0	1	0
Percentage	82	87	14	13	0	0	4	0

Table 14. Summary of the results (Score 1 and 2 considered successful, score 3 and 4 failure)

Figures 9-10 show examples of the X-ray examination of orthograde and retrograde treatments.

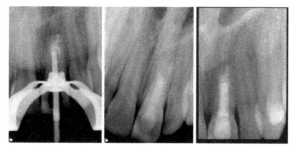

Fig. 9. Tooth 21 (patient14) a) condensing with a Gutta-percha pointer, b) just after treatment and c) at two year control

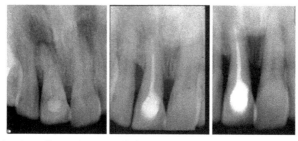

Fig. 10. Tooth 21 (patient 9) at treatment (left) and at two year control (middle) and at 5 year control (right)

In summary 21 out of 22 treated teeth have acceptable results being either symptom free or judged healed after clinical examination. The single failure can probably not be attributed to the material, but rather to the difficulty of treating and sealing a multi-channelled tooth. The use of CA's as root canal sealers is indirectly supported in "Introduction to Dental Materials" by van Noort (van Noort, 1994), where the following materials characteristics are looked for; biocompatible, dimensionally stable, antibacterial and bioactive. The results in this study can be interpreted as a success in meeting these materials requirements.

Already in the 1970s, Calcium aluminate (CA) was suggested as a biomaterial and tested *in vivo*. Hentrich et al (Hentricht et al, 1994) compared CA with alumina and zirconia in an evaluation of how the different ceramics influenced the rate of new bone formation in femurs of rhesus monkeys. Hamner et al (Hamnar and Gruelich, 1972) presented a study in which 22 CA roots were implanted into fresh natural tooth extraction sites in 10 baboons for periods ranging from 2 weeks to 10 months. In both studies CA successfully met the criteria for tissue adherence and host acceptance.

4.1.3 Dental filling materials

An important feature of the hydration mechanisms of the Ca-aluminate based materials is the nanostructural integration with and the high shear strength developed towards dental tissue. This makes both undercut (retention) technique and bonding techniques redundant. The Ca-aluminate approach to dental filling technique is new. With this technique, the chemical reactions cause integration when the bioceramic material is placed in the oral cavity at body temperature and in a moist treatment field. Figure 11 shows a TEM (transmission electron microscopy) illustration of the interface between the CA-based material and dentine. This establishes a durable seal between bioceramic and tooth. Whereas amalgam attaches to the tooth by mechanical retention and resin-based materials attach by adhesion, using bonding agents, etchants, light-curing or other complementary techniques, the CA-materials integrate with the tooth without any of these, delivering a quicker, simpler and more robust solution.

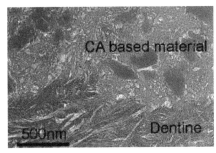

Fig. 11. Nanostructural integration of CAPH-material with dentine (gray particles in the biomaterial are glass particles)

The general aspects of Ca-aluminate based materials have been presented in two Ph D Thesis-publications. Important aspects of Ca-aluminate materials as dental filling materials are dealt with, such as dimensional stability, acid corrosion and wear resistance, and biocompatibility and mechanical properties (Kraft 2002, Lööf et al 2003).

4.1.4 Coatings on dental implant and augmentation

For successful implantation of implants in bone tissue, early stabilisation is of great importance (Ellingsen and Lyngstaadas, 2003). This includes both orthopaedic and dental implants. Even small gaps may lead to relative micro-motions between implant and the tissue, which increases the risk of implant loosening over time due to formation of zones of fibrous tissues at the implant-tissue interface. Early loading of implants is of particular interest for dental implants (Vercaigne et al, 1998). The use of surface coatings technology is today an established method to reduce the problem with poor interfacial stability for implants. With coatings technology, structural characteristics of the implant (e.g. strength, ductility, low weight or machinability) may be combined with surface properties promoting tissue integration. There are several established coating deposition techniques, e.g. physical vapour deposition (sputtering) and thermal spraying techniques. Coatings based on calcium phosphates are the most used ones.

This section deals with coatings deposited with established methods, with the aim of improving particularly the early stage anchoring of metal implants to bone tissue by exploring *in vivo* hydration of coatings or pastes based on chemically curing ceramics. The study focuses on calcium aluminate in the form of coatings and paste. Results are presented from an implantation study with flame-sprayed coating on titanium implants and uncoated implants augmented with a calcium aluminate paste in the hind legs of rabbits. Implants were applied with the paste composed of a mixture of $CaO \cdot Al_2O_3$ and $CaO \cdot 2Al_2O_3$. The paste was applied manually as a thin layer on the threaded part of the implant just before implantation. The uncoated and coated implants were sterilised with hot dry air at 180 °C for 2 hrs. Female albino adult New Zealand White rabbits with a body weight around 2.5 kg were used. Each animal received four implants, two in each hind leg. Implants were placed in the distal femoral metaphysis as well as in the proximal tibial metaphysis. Surgery followed standard procedure. The implants were screwed into predrilled and threaded cavities. Necropsy took place after 24 hrs, 2 and 6 weeks (Axen et al, 2005).

No negative effects of the implants on the general welfare of the animals were observed. The healing progressed in a normal and favourable way. As for the removal torque recordings, all calcium aluminate coatings types provided an improved implant anchoring to bone tissue after *in vivo* hydration, as compared to that of the pure metal implants. Implants on the tibia and femur side of the knee gave similar removal torques. Table 15 provides average values from both tibia and femur sides.

Implant type	24 hrs	(n)	2 weeks	(n)	6 weeks	(n)
Flame spraying	7.0	(8)	7.0	(8)	25	(6)
Paste augmentation	6.6	(8)	15	(6)	13	(4)
Rf-PVD	12	(4)	-	-	10	(4)
Uncoated reference	3.8	(8)	5.7	(6)	14	(4)

Table 15. Removal torque (Ncm) for dental implants in rabbit hind legs (tibia and femur).

24 hrs after implantation, calcium aluminate in-between the implant and tissue increased the removal torque to about double that of the uncoated reference implants, independently of means of application (coatings or paste). This is considered to be attributable to the point-welding according to integration mechanism 6 above. Two weeks after implantation, implants combined with paste augmentation provide the highest removal torque; flame sprayed coatings also improve the torque relative to the uncoated system. Six weeks after implantation, all systems are relatively similar (considering the uncertainty due to scatter and statistics), apart from the sprayed system which shows significantly higher values.

Fig. 12. High-resolution TEM of Ti – CA-paste interface, nano-mechanical integration, bar = 10 nm.

4.2 Orthopaedic applications

Within orthopedics the following areas for Ca-aluminate based materials have been identified; percutaneous vertebroplasty (PVP) and kyphoplasty (KVP), trauma and general augmentation.

The benefits of the injectable ceramic biomaterials based on CA related to orthopaedic applications are

During the surgical procedure (Engqvist et al, 2005, Lööf et al 2008)

- High radiopacity allows for superior visibility of the cement during and increases the probability to detect potential leakages during injection (See Figs. 13-14 below)

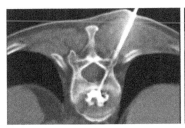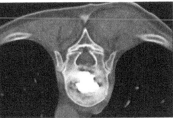

Fig. 13. Percutaneous vertebroplasty using a CA-material.

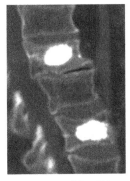

Fig. 14. Vertebral compression factures, restored by CA-material.

- High and linearly increasing viscosity reduces the risk of leakage and gives a
- predictable handling.
- High cohesiveness optimizes the cement's filling pattern in the vertebra.
- No toxic or smelling fumes

And after the procedure (Jarmar et al, 2008, Engqvist et al 2006)

- Mechanical strength
- Biocompatibility including integration
- Long-term stability i.e. non-resorbable systems.

4.3 Drug carrier for drug delivery

General aspects of ceramics for use in drug delivery of drugs are presented by Ravaglioli et al and by Lasserre and Bajpaj (Ravaglioli et al, 2000, Lasserre and Bajpaj, 1998). A short description of carrier materials for drug delivery using chemically bonded ceramics, especially Ca-aluminate and/or Ca-silicate systems are given below. The CBC carrier material based on CA and CS structures exhibit some attractive features. The manufacturing procedure at low temperatures, where no or limited degradation of the medicaments occur, and the microstructure developed with open porosity as nano-sized channels as described above, are the basic features that open up a possibility for controlled release of medical agents. The precursor powder cures as a result of hydration reactions, between a ceramic oxide powder, primarily Ca-silicates and/or Ca-aluminates, and water. Through the hydration, new phases of hydrates are formed, which to a great part establish the microstructures needed to control the release of drugs incorporated in the injectable precursor material. An injectable material is formed into a paste by mixing it with a water-based hydration liquid, which is then ready to be injected. Directly after the injection, the paste starts to develop the final microstructure. The water-based liquid may also comprise viscosity-controlling additives. These may be loaded with the drug before preparation of the final injectable paste. A couple of unique reaction conditions related to the production of materials yields materials with a variety of possible microstructures with porosities from the nanoscale to the microscale. variety of possible microstructures with porosities from the nanoscale to the microscale. 3) pore size and pore channel size, and 4) combination of different porosity structures (Hermansson, 2010). Porosity generated during the hydration of the Ca-aluminates and Ca-silicates is open porosity due to the reaction mechanism, and can be in the interval of 5-60 vol.-%. The average pore channel size (i.e. the diameter of the pores formed between the particles of the hydrated material) may be 1-10 nm. The crystal size of the reacted hydrates is in the interval 10-50 nm. This was established by BET-measurements, where the specific surface area of dried hydrated CA was determined to be in the interval 400-500 m², corresponding to a particle size of approximately 25 nm, and by HRTEM [7], Fig. 1 below. When short hydration time and/or low amount of water, or moisture at relative humidity > 70 %, are used, additional porosity is achieved with pore sizes in the interval 0.1-1 micrometer due to incomplete reaction. The different pore sizes obtained can be utilized for controlled release of drugs, when the Ca-aluminate implant material also works as a carrier of medicaments.

Drug loading and controlled release of drugs

The following properties are of significance with regard to the carrier for controlling the drug release; Type of ceramic precursor for producing the chemically bonded ceramic, grain size distribution of the precursor powder particles and general microstructure of the material, he microstructure of the additional particles for drug incorporation, and additives to ensure complementary porosity.

The loading of the drug can be performed in several ways. The drug may be included, either partially or fully, in the powder or in the hydration liquid. Time and temperature for hydration are selected with regard to the drug and drug loading and to the selected release criteria. The manufacturing of the carrier can be done completely before or during loading of the drug. This renders a controlled release time to be selected from a few hours to days and months.

The drug is introduced in the carrier by mixing the drug into the precursor powder, or the hydrated CBCs or other porous phases. The material can be formed into a paste by mixing it with a water-based hydration liquid. The powder can also be pressed into pellets, which thereafter are soaked in the liquid. The paste or the soaked pellets start to develop the microstructure that to a great extent will contribute to the controlled release of the drug. The time and temperature after the mixing will determine the degree of hydration, i.e. the porosity obtained. The porosity can be controlled within a broad interval of open porosity.

5. Conclusion

The Ca-aluminate technology provides a platform upon which Ca-aluminate based materials may work as a general biomaterial and as a complement to other chemically bonded ceramics based on phosphates, silicates or sulphates. Identified areas are in the first place within the dental and orthopedic areas, where injectable stable biomaterials are required. These include also properties as bioactivity related to apatite formation, antibacterial properties as well as nanostructural features useful for carriers for controlled drug delivery. The studies presented in this paper can be summarised as follows;

- Nano-structural integration and apatite formation provide important benefits to both the dentist and patient, notably minimal micro-leakage, perfect seal at the interface between tooth and material and as a result longer-lasting treatment results.
- The potential use of the Ca-aluminate materials for implant applications is based on the following features: nanostructural integration with tissue, possible apatite formation, and a mass increase yielding early point welding between the biomaterial and surrounding.
- The following product areas have been identified based on experimental material data, pre-clinical studies, pilot studies and on-going clinical studies: dental cement, endodontic products (orthograde and retrograde), sealants, restoratives, underfillings, and pastes for augmentation and dental implant coatings.
- Consequences of nanostructural contact integration of the Ca-aluminate system are reduced risk of secondary caries and restoration failure, and reduced post-operative sensitivity.
- The Ca-aluminate material can be used as a vehicle for transport and delivery of the medicament and as an injectable implant. The combination of the material as carrier and implant material makes site-specific placement of drugs and implants possible. By introducing optional additives, or by changing the w/c ratio, the release time can be controlled from short time periods (a few hours) to prolonged time periods (day and weeks). The release time is also dependant upon where the drug is placed. In cortical bone a release time of months seems possible.

- The Ca-aluminate materials are not degradable and do not induce clotting or haemolysis.
- The first product in a series - a dental luting cement- was recently launched on the European and the US markets.

6. Acknowledgment

Results presented in this chapter are mainly based on two decades of research within Doxa AB and the Eng. Sci. Dept., The Angstrom Laboratory, Uppsala Univesity, Sweden.

7. References

Mangabhai, R. J. (1990), Calcium Aluminate Cements, Conf proceedings, Chapman and Hall

Hench, L (1998) Biomaterials: a forecast for the future, *Biomaterials* Vol 19 1419-1423

Nilsson, M. (2003) *Ph D Thesis*, Injectable calcium sulphates and calcium phosphates as bone substitutes, Lund University

Scrivener, K.L.and Capmas A.; (1998) Calcium aluminate cements, In: Lea's *Chemistry of Cement and Concrete*, Ed. Hewett P.C., Arnold: Paris, p 709-771

Kraft, L. (2002) *Ph D Thesis*, Calcium aluminate based cements as dental restorative materials. Faculty of Science and technology, Uppsala University, Sweden

Loof, J; Engqvist, H.; Lindqvist, K.; Ahnfelt, N-O.; Hermansson L; (2003), Mechanical properties of a permanent dental restorative material based on calcium aluminate, *Journal of Materials Science: Materials in Medicine*, 14, No. 12 1033-1037

Engqvist, H.; Edlund S.; Gomez-Ortega, G.; Loof, J.; and Hermansson, L.; (2006) In vitro mechanical properties of a calcium silicate based bone void filler, *Key Eng. Mater.* 361-363 (369-376)

Muan, A.; Osbourne, E. A.; (1965) Phase equilibria among oxides. Adison-Wesley; New York

Lewis, G.; (2006) *J. Biomed. Matls. Res. PartB: Applied Biomaterials* 76B: 456-468

H. Engqvist, J. Loof, S. Uppstrom, M. W. Phaneuf, J. C. Jonsson, L. Hermansson, N-O. Ahnfelt, (2004), Transmittance of a bioceramic calcium aluminate based dental restorative material, *Journal of Biomedical materials Research Part B: Applied Biomaterials* vol. 69 no. 1, 94-98

Lööf, J. (2008) *Ph D Thesis*, Calcium-aluminate as biomaterial: Synthesis, design and evaluation. Faculty of Science and Technology, Uppsala, University, Sweden

Lööf, J.; Engqvist H.; Gómez-Ortega, G.; Spengler, H.; Ahnfelt, N-O.; Hermansson, L.; (2005) Mechanical property aspects of a biomineral based dental restorative system, *Key Engineering Materials* Vols. 284-286, 741-744

Loof, J.; Engqvist, H.; Hermansson, L.; Ahnfelt, N-O. (2004) Mechanical testing of chemically bonded bioactive ceramic materials, *Key Engineering Materials* Vols. 254-256, 51-54

Hermansson, L.; Kraft, L.; Lindqvist, K.; Ahnfelt, N-O.; Engqvist, H.; (2008) Flexural Strength Measurement of Ceramic Dental Restorative Materials, *Key Engineering Materials*, Vols. 361-363, 873-876

Williams, D.D.; (1987) Definitions in Biomaterials, Chaper 8, pp 66-71, Ed. Elsevier, New York

Cao, W.; Hench L.L.(1996), Bioactive materials, *Ceramics International*, Vol 22 493-507 ISO 10993:2003 EN 29917:1994/ISO 9917: 1991

Liu, Y.; Sahlberg, L.; Kraft, L.; Ahnfelt, N-O.; Hermansson, L.; Ekstrand, J. (2002) Aspects of Biocompatibility and Chemical Stability of Calcium-Aluminate-Hydrate Based

Dental Restorative Material, Paper IX in *Ph D Thesis* by L. Kraft, Uppsala University 2002

Engqvist, H.; Lööf, J.; Kraft, L.; Hermansson, L.; (2004) Apatite formation on a biomaterial-based dental filling material, *Ceramic Transactions,* Vol 164 , Biomaterials: Materials and Applications, V. 2004;164:37-42.

Faris, A.; Engqvist, H.; Lööf, J.; Ottosson, M.; Hermansson, L.; (2006) In vitro bioactivity of injectable ceramic orthopaedic cements, *Key Eng. Mater.* 309-11, 1401-1404

Engqvist, H.; Couillard, M.; Botton, G.A.; Phaneuf, M.P.; Axén, N.; Ahnfelt, N-O. and Hermansson, L.; (2005), In vivo bioactivity of a novel mineral based based orthopaedic biocement, *Trends in Biomaterials and Artificial Organs*, 19, 27-32

Engqvist, H.; Schultz-Walz, J-E.; Lööf, J.; Bottom, G.A.; Mayer, D.; Phaneuf, M.W.; Ahnfelt, N-O.; Hermansson, L.; (2004), *Biomaterials* Vol 25, 2781-2787

FAO/WHO Joint Expert Committee on Food Additives (JECFA)

Jefferies, S.R.; Appleby, D.; Boston, D.; Pamiejer, C.M..; and Lööf, J.; (2009). Clinical performance of a bioactive dental luting cement- A prospective clinical pilot study. *J. of Clinical Dentristy* XX, No. 7, 231-237 ISO 10993-12: 1992 ISO 10993-5:1992, clause 8.4.1

Tang, A.T.H.; Li, J.; Ekstrand, J.; and Liu, Y. (2002), Cytotoxicity tests of in situ polymerized resins: Methodological comparisons and introduction of a tissue culture insert as a testing device *J Biomed Mater Res*, 45, 214-222.

Schmalz, G.; Rietz, M.; Federin, M.; Hiller, H-A.; Schweikl, H.; (2002) Pulp derived cell response to an inorganic direct filling material, Abstact, presented at Cardiff Conference,

Hermansson, L.; Höglund, U.; Olaisson, E.; Thomsen, P.; Engqvist, H.; (2008) Comparative study of the bevaiour of a novel injectable bioceramic in sheep vertebrae, *Trends in Biomater. Artif. Organs*, Vol 22, 134-139

Axén, N.; Ahnfelt, N-O.; Persson, T.; Hermansson, L.; Sanchez, J. and Larsson, R.; (2004), A comparative evaluation of orthopaedic cements in human whole blood, *Proc. 9th Ceramics: Cells and tissue,* Faenza

Hermansson, L.; Lööf J. and Jarmar, T.; (2009) Integration mechanisms towards hard tissue of Ca-aluminate based materials, *Key Eng. Mater.* Vol 396-398, 183-186

Hermansson, L.; Engqvist, H.; Lööf, J.; Gómez-Ortega G. and Björklund, K.; (2006) Nano-size biomaterials based on Ca-aluminate Advances in Science and technology, *Key Eng. Mater.* Vol 49 21-26

Hermansson, L.; Lööf, J. and Jarmar, T.; (2008) Injectable ceramics as biomaterials today and tomorrow, in *Proc. ICC 2,* Verona 2008

Ravagliolo, A.; and Krajewski, A.; (1992) *Bioceramics,* Chapman and Hall, Mjör, I.; (2000) *Int Dental Journal*, Vol 50 [6], 50

Pameijer, C.H.; Jeffries, S.R.; Lööf, J. and Hermansson, L.; (2008) Microleakage Evaluation of XeraCem in Cemented Crowns, *J Dent Res.* 2008;87(B):3098.

Pameijer, C. H.; Jeffries, S.R; Lööf J. and Hermansson, L.; (2008) Physical properties of XeraCem, *J Dent Res.* 2008; 87(B):3100.

Pameijer, C. H.; Jeffries, S.R.; Lööf, J.; and Hermansson, L.; (2008) A comparative crown retention test using XeraCem, *J Dent Res.* 2008; 87 (B):3099.

Jefferies, S.R.; Pameijer, C.H.; Appleby, D.; Boston (2009), One month and six month clinical performance of XeraCem®. *J Dent Res.* 2009;88(A):3146

Jefferies, S.R.; Pameijer, C.H.; Appleby, D.; Boston D.; Lööf, J.; Glantz, P-O.; (2009), One year clinical performance and post-opreative sensitivity of a bioactive dental luting cement, *Swed. Dent. J.* 2009 Vol 33 193-199

Hermansson, L.; Faris, A.; Gomez-Ortega, G.; Abrahamsson, E:; Lööf, J.; (2010), Calcium aluminate based dental luting cement with improved properties – an overview, Advances in Bioceramics and Porous Ceramics, 34th Int Conf on Advanced Ceramics and Composites,, Ed. Wiley, *Ceramic Eng. And Sci. Proc, Vol 31, p 27-38*

Jefferies, SR.; Pameijer, CH.; Appleby, D.; Boston, D.; Lööf, J.; Glantz, P-O.; (2009) One year clinical performance and post-operative sensitivity of a bioactive dental luting cement – A prospective clinical study. *Swed Dent J.* 2009; 33:193-199.

Haumann, C.H.J. and R.M. Love, (2003) Biocompatibility of dental materials used in contemporary endodontic therapy: a review. Part 2, Root–canal–filling materials, *International Endo J*, 36: p. 147-160.

Niederman, R. and J.N. Theodosopoulou, Review: A systematic review of in-vivo retrograde obturation materials. *Int Endo J*, 2003. 36: p. 577-585.

Alamo, H.L., et al., A Comparison of MTA, Super-EBA, composite and amalgam as root-end filling materials using a bacterial microleakage model, *International Endodontic Journal*, 1999. 32: p. 197-203.

Kraft, L.; Saksi, M.; Hermansson, L.; Pameijer, CH.; (2009) A five-year retrospective clinical study of a calcium-aluminate in retrograde endodontics. *J Dent Res.* 2009; 88(A):1333.

Noort, R.v., (1994), *Introduction to Dental Materials.* Mosby Hentricht, R.L; (1971), An evaluation of inert and resorbable ceramics for future clinical applications, *J. Biom. Res.* 5(1): 25-51

Hamner, J.E., Reed, M.; and Gruelich, R.C.; Ceramic root implantation in baboons. *J. Biom. Res*, 1972 6 (4): p. 1-13.

Ellingsen, J-E.;Lyngstadaas, S.P. ; (2003), *Bioimplant interface, improving biomaterials and tissue reactions*, CRC Press LLC. Vercaigne S. et al, (1998) Bone healing capacity of titanium plasma-sprayed and hydroxylapatite coated oral implants, *Clin. Oral Implants Res*, 9, 261

Axén, N.; Engqvist, H.; Lööf, J.; Thomsen P.; and Hermansson, L.; (2005) , In vivo hydrating calcium aluminate coatings for anchoring of metal implants in bone, *Key Eng. Mater.* Vols. 284-286 831-834 21

Jarmar, T.; Uhlin, T.; Höglund, U.; Thomsen, P.; Hermansson, L.; and Engqvist, H.; Injectable bone cements for Vertebroplasty studied in sheep vertebrae with electron microscopy, *Key Engineering Materials* Vols. 361-363

Engqvist, H.; Hermansson, L.; (2006) Chemically bonded nano-size bioceramics based on Ca-aluminates and silicates, *Ceramic Transactions.* 2006; 172: 221-228. Ravaglioli et al, (2000), *J Mater Sci Mater Med.* 2000 11(12):763-7

Lasserre; and Bajpaj, (1998) Critical Reviews in *Therapeutic Drug Carrier Systems,* Vol 15, 1

Hermansson, L.; (2010), Chemically bonded bioceramic carrier systems for drug delivery, Advances in Bioceramics and Porous Ceramics, Ed. Wiley, 34th Int. Conference on Advanced Ceramics and Composites, *Ceramic Eng. and Sci. Proc,* Vol 31, p 77-88

3

Ulvan: A Versatile Platform of Biomaterials from Renewable Resources

Federica Chiellini and Andrea Morelli
Laboratory of Bioactive Polymeric Materials for Biomedical and Environmental Applications (BIOlab) UdR-INSTM – Department of Chemistry and Industrial Chemistry, University of Pisa
Italy

1. Introduction

Biomass represents an abundant renewable resource for the production of bioenergy and biomaterials and its exploitation could lead to overcome the dependence from petroleum resources. Indeed fossil energy and chemical sources are not unlimited and there is a critical need to turn the current way of life back to a sustainable manner. The conversion of biomasses into high value chemicals, energy and materials is nowadays gaining more and more attention and represents the final goal of the "Industrial Biorefinering". Indeed Biorefinery aims at the optimum exploitation of biomass resources for the production of materials that eventually might replace the conventional products from fossil/non renewable resources, thus decisively contributing to the development of a sustainable system. The great challenge in which Biorefinering is involved is the possibility of creating high value products from low value biomasses. In this view, the feasibility of using starting materials obtainable from organic waste sources (agricultural, municipal and industrial waste) or having harmful effects on the environment (algae) as feedstock can represent the strategy of election for the production of sustainable materials.

To this aim algae could represent a potentially advantageous biomass to be explored since they are very abundant and cheap and very often involved in uncontrolled proliferation processes detrimental for marine and aquatic environments (Barghini et al., 2010, Chiellini et al., 2008, 2009, Fletcher, 1996). Today most of the naturally produced and harvested algal biomass is an unused resource and often is left to decompose on the shore creating waste problems (Morand et al., 2006). The current use of this huge underexploited biomass is mainly limited to food consumption and as bio-fertilizer, but its potentiality as renewable and sustainable feedstock for energy and material production is gaining more and more attention (Demirbas A. & Demirbas M.F., 2011). Indeed microalgae have been considered to be an excellent source for biodiesel production since are characterized by high growth rates and high population densities, ideal for intensive agriculture and may contain huge lipid amounts, needed for fuel production (Christi, 2007). Besides biodiesel, algae can be cultivated and can be used as a feedstock for the production of bioethanol (John et al., 2011). In particular macroalgae (seaweed) can produce huge amount of carbohydrates per year

(Matsumoto et al., 2003) that suitably processed through specific fermentation processes would provide renewable and sustainable biofuel.

Algae represent also an advantageous resource of chemicals and building block materials that can be tailored through proper biorefinering processes according to the different envisaged applications. The rising demand for natural instead of synthetic materials especially in biomedical applications where high biocompatibility and no adverse effects for the host organism are required (Mano et al., 2007), has led to an outburst of scientific papers involved in the study of biobased materials. Among these, polysaccharides could represent the best candidate since abundant, biocompatible and displaying a pronounced chemical versatility given by the great number of chemical functionalities present in their structures. The list of known natural carbohydrates is continuously growing, owing to new discoveries in animal and plant material (Tsai, 2007). They can be used in their native form or after proper chemical modifications made according to the final applications (d'Ayala et al., 2008). The use of polysaccharides of animal origin (e.g. heparin and hyaluronic acid) in biomedical applications is not straightforward since it can raise concerns about immunogenicity and risk of disease transmission (Stevens, 2008) Indeed these materials require very accurate purification treatments aimed to free them from biological contaminants, in contrary to polysaccharides of plant (e.g. cellulose and starch) or algal origin (e.g. alginate). Polysaccharides of algal origins are gaining particular attention due to their abundance, renewability (Matsumoto et al., 2003) and to their peculiar chemical composition not found in any other organisms. Over the last few years medical and pharmaceutical industries have shown an increasing interest in alginate (d'Ayala et al., 2008), an anionic polysaccharide widely distributed in the cell walls of brown algae. This biopolymer has been largely used for its gel forming properties. Due to its non-toxicity, unique tissue compatibility, and biodegradability, alginate has been studied extensively in tissue engineering, including the regeneration of skin (Hashimoto et al., 2004), cartilage (Bouhadir et al., 2001), bone (Alsberg et al., 2001), liver (Chung et al., 2002) and cardiac tissue (Dar et al., 2002).

A very intriguing feature that distinguishes algal biomass from other resources is that it contains large amounts of sulphated polysaccharides, whose beneficial biological properties (Wijesekara et al., 2011) prompt scientists to increase their use in the biomedical fields. Indeed the presence and the distribution of sulphate groups in these polysaccharides are reported to play an important role in the antiviral (Damonte et al., 2004), anticoagulant (Melo et al., 2004), antioxidant (Rocha de Souza et al., 2007) and anticancer (Athukorala et al., 2009) activity of these materials.

The chemical composition of the sulphated polysaccharides extracted from algae, including the degree and the distribution of the sulphate groups, varies according to the species, and the ecophysiological origin of the algal sources (Rioux et al., 2007). Anyhow, a structural differentiation depending on the different taxonomic classification of the algal origin, has been found. According to the mentioned classification the major sulphated polysaccharides found in marine algae include fucoidan from brown algae, carrageenan from red algae and ulvan obtained from green algae (Figure 1).

Ulvan polysaccharides possess unique structural properties since the repeating unit shares chemical affinity with glycoaminoglycan such as hyaluronan and chondroitin sulphate due to its content of glucuronic acid and sulphate (Figure 2).

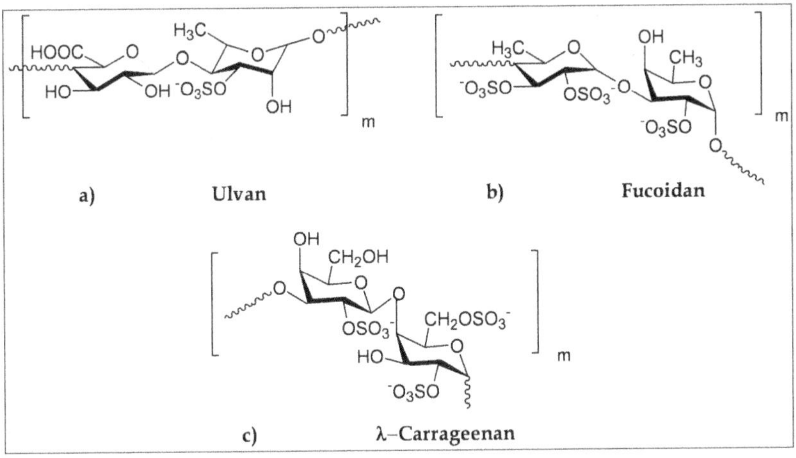

Fig. 1. Chemical structure of the dimeric repeating unit of a) Ulvan, b) Fucoidan, c) λ - Carrageenan.

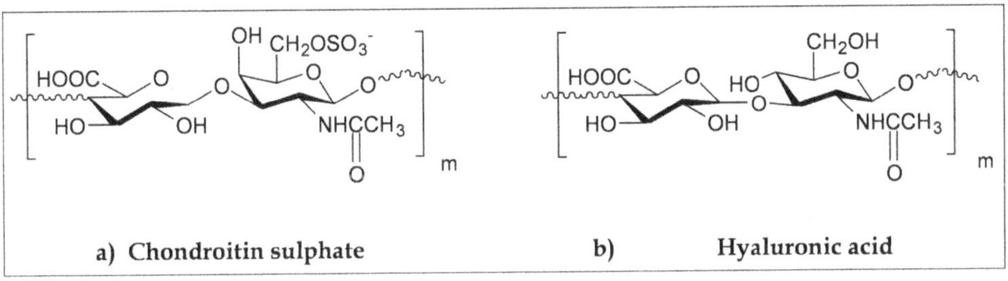

Fig. 2. Chemical structure of the dimeric repeating unit of a) Chondroitin sulphate, b) Hyaluronic acid.

This resemblance and the possibility of obtaining this material from cheap and renewable resources make it worthwhile a deeper investigation on the biological activity and the feasibility of using this polysaccharide for biomedical applications.

2. Ulvan properties

Ulvales (Chlorophyta) are very common seaweeds distributed worldwide. The two main *genera Ulva* and *Enteromorpha* are sadly known for being involved in processes detrimental for the aquatic environment. Indeed this algal biomass proliferates very quickly in eutrophic coastal and lagoon waters in the form of "green tides" leading up to hypoxia and death of most of aquatic organisms (Morand & Brian, 1996). Environmental concerns arise also for the disposal of this huge biomass that is mostly left do degrade on the shore creating nuisance problems, so that its exploitation could represent a remedy to related environmental and economical concerns.

To date, this biomass has very low added value and its use is limited to food consumption (Bobin-Dudigeon et al., 1997) composting (Mazè et al., 1993) and methane production

(Brand & Morand, 1997) but as it will be stressed in the following part of the chapter, the chemicals and polymers of this underexploited biomass along with their abundance, biological properties and "renewability" represent a potential source to be explored.

2.1 Chemical-physical properties
2.1.1 Ulvan composition

Green algae such as *Ulva sp.* are known to contain high amounts of good-quality protein, carbohydrate, vitamins and minerals (Taboada et al., 2010). Among these, polysaccharides are gaining increasing attention as they possess unique physical and chemical properties representing a versatile material platform for potential biological applications.

Ulvan represents a class of sulphated heteropolysaccharide extracted from the cell wall of green seaweeds belonging to *Ulva sp.* whose composition has been extensively debated (Lahaye & Robic, 2007; Robic et al., 2009;) and showed to vary according to several factors including the period of collection, the ecophysiological growth conditions, the taxonomic origins and the post-collection treatment of the algal sources (Lahaye & Robic, 2007).

Four types of polysaccharides are reported to be contained in the biomass of *Ulva sp.*, including the water soluble Ulvan and insoluble cellulose as major one and an alkali-soluble linear xyloglucan and glucuronan in minor amounts (Lahaye & Robic, 2007). Ulvan represents the major biopolymeric fraction of the cell wall having the function of maintaining the osmolar stability and protection of the cell (Paradossi et al., 2002). As usually found in polysaccharides present into the cell walls, Ulvan is present in close association with proteins and the conventional methods of extraction and purification resulted not completely effective in the removal of the protein fraction even after a specific deproteinization protocol (Alves et al., 2010).

Extraction is conventionally achieved by using warm water solution (80-90°C) containing ammonium oxalate as divalent cation chelator and the recovery of Ulvan is generally obtained by precipitation in ethanol. The yield of extraction usually ranges from 8% to 29% of the algal dry weight depending on the applied purification procedure (Lahaye & Axelos, 1993; Lahaye et al., 1994).

The sugar composition of Ulvan is extremely variable but rhamnose, xylose, glucuronic and iduronic acid and the presence of sulphate groups have been identified as the main constituents of the polymer (Paradossi et al., 2002; Robic et al., 2009). These monomers are arranged in an essentially linear fashion even though a slight degree of branching has been found (Lahaye & Robic, 2007). The chemical heterogeneity of Ulvan is partially striken by a "structural motif" found within the heteropolymer chain essentially given by the presence of repeating dimeric sequences constituted by aldobiuronic acid disaccharides designated as type A (glucurorhamnose 3-sulphate, A_{3s}) and type B (iduronorhamnose 3-sulphate, B_{3s}) (Figure 3).

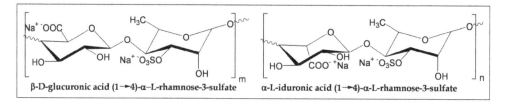

β-D-glucuronic acid (1→4)-α-L-rhamnose-3-sulfate   α-L-iduronic acid (1→4)-α-L-rhamnose-3-sulfate

Fig. 3 Structure of the main disaccharide repeating units in Ulvan.

The most striking feature that distinguishes the chemical composition of Ulvan from that of the other polysaccharides of marine origin is, therefore, the presence of uncommon sugar such as iduronic and sulphated rhamnose displaying a close similarity with mammalian glycosaminoglycans. To this view Ulvan and related polysaccharides could represent an abundant and cheap feedstock for the substitution of heparinoid substances commonly used in biomedical applications solving the problems related to their isolation and purification (Alban et al., 2002).

2.1.2 Ulvan conformation

The physical properties of polymeric materials are deeply affected by the association and conformation assumed by the constituting chains in the final product. The balance between ordered crystalline and disordered amorphous structures dictates the ultimate mechanical properties of the polymeric material. Indeed the possibility of forming crystalline regions inside a polymeric structure could even generate physical crosslinks between the chains inducing ultimately to the formation of stiff networks, as in the case of polyvinyl alcohol (Ricciardi et al, 2005). The achievement of suitable mechanical properties for a material to be used in biomedical applications, namely tissue engineering, represent a key requirement to fulfil since the final product must provide a physical support for the cell growth and differentiation.

Past investigations on this issue revealed an essentially disordered conformation of Ulvan (Paradossi et al., 1999) mainly induced by the heterogeneous chemical composition of this polysaccharide. The local regularity given by the repeating aldobiuronic units, denominated as A_{3s} and B_{3s} (Figure 3), is believed to be sufficient for the formation of transient "junction-zones" responsible for the formation of the weak gel that ulvan is known to perform in nature (Paradossi et al., 2002). The stability of these ordered structures can be affected by the attractive and repulsive interactions that form between the functional groups of the polysaccharide, and in particular by the electrostatic forces. Ulvan is an anionic polyelectrolyte as it contains carboxylic and sulphate groups inside its structure, so that its net charge strongly depends on the pH and ionic strength of the working medium. The net charge on Ulvan is found to affect the conformation of its polymeric chains and ultimately controls the order to disorder transitions given by the locally regular sequences (Paradossi et al., 2002). The conformational change from an ordered structure present in the uncharged chain, i.e. the protonated form of ulvan, toward a disordered state, happens when a critical charge density is reached and is induced only in the chemically regular portions of the chains. The structures of the ordered sequences have been hypothesized on the basis of molecular modelling calculations and are compatible with the formation of helical conformations inside homogeneous portions of the chains containing the repeating units A_{3s} and B_{3s} (Paradossi et al., 2002).

The presence of ordered structures limited only in the regular sequences of the Ulvan polymeric chains is not sufficient to provide enough "junction-zones" for the preparation of a material with mechanical properties suitable for biomedical applications. For this purpose Ulvan has to be modified through the introduction of chemical groups or molecules that increase the number of "junction-zones".

2.1.3 Ulvan morphology and solubility

The possibility of chemically modifying Ulvan is strongly dependent on the physical availability of its functional groups so that its solubility and morphology in the working medium could affect deeply its reactivity. Ulvan has been shown to dissolve only in water due to its charged and highly hydrophilic nature. Nevertheless, the obtained solutions are not transparent, indicating the formation of microaggregates of polymeric material not fully dispersed in the solvent. Indeed TEM analysis of Ulvan revealed the presence of aggregates of spherical shaped forms partially linked by strands-like filaments (Robic et al., 2009). This necklace-like ultrastructure is usually formed by polyelectrolyte material in poor solvent conditions (Dobrynin, 2008) so that even water can not be considered a good solvent for Ulvan. The large presence of methyl groups provided by the rhamnose repeating unit has been considered responsible for the unusual hydrophobic behavior of this highly charged polysaccharide (Robic et al., 2009).

The unusual low intrinsic viscosity of Ulvan in solution can also be ascribed to the presence of condensed spherical shaped aggregates not typical for polyelectrolytes whose conformation usually expands in the form of charged filaments and leads to an increase in the viscosity (Dobrynin et al., 1995). The formation of microaggregates in solution does not allow also a reliable mass analysis of Ulvan, whose different type of aggregation affects deeply the peak distributions usually found on the GPC chromatograms (Robic et al., 2009). Being a polyelectrolyte, both the ionic strength and the pH of the dissolving medium would affect the solubility and the morphology of Ulvan. Indeed the association of the bead-like aggregates in a necklace-type ultrastructure is promoted by the ionic interactions of carboxylated groups as demonstrated by its rupture at pH below the pKa of glucuronic acid (3.28) (Robic et al. 2009). In basic conditions (pH 13) the bead-like structures resulted to collapse into a dense homogeneous network likely prompted by the ionic interactions of carboxylate and sulphate groups. The type and amount of counter-ion in solution could also contribute to chain expansion or condensation as demonstrated by the aggregative propensity of Ulvan at low NaCl concentration observed by light scattering and rheological measurements (Lahaye & Robic, 2007).

The tendency of Ulvan to form aggregates in aqueous solution and its insolubility in almost every organic solvents limit the number of functional groups available for chemical modifications thus hampering its potential versatility. But its great number of reactive groups still present on the "free" surface exposed outside the aggregate and the possibility to optimize the solvent variables (pH and ionic strength) that affect the dispersion of the polymer in solution make Ulvan a suitable reactive platform, tailorable according to the envisaged application.

2.2 Biological activity

The possibility of using bio-based materials in almost every technological field and particularly in biomedical applications is challenging and can be considered the strategy of election for limiting environmental concerns and create a virtuous circle of sustainability.

Biomaterials possess the essential prerequisite of renewability and biocompatibility and as such are worth of deep investigations as main candidates for the substitution of synthetic petroleum-based materials, well known for being not renewable and often not biocompatible.

Biodegradability represents also an important property possessed by biomaterials and it is especially required in materials used for biomedical applications with specific reference to tissue engineering and regenerative medicine. Not only the material has to be safe but also the products of degradation should be non-toxic and easily cleared from the body. Biomaterials that other than being renewable, biocompatible and biodegradable are able to induce a beneficial biological activity on the host organism or on the environment can be considered even more intriguing. This may be the case of Ulvan. Most of the positive health effects induced by this polysaccharide are generated by the presence of sulphate groups in its structure (Wijesekara et al., 2011). A wide list of beneficial biological effects reported by the literature span from antioxidant (Qi et al., 2006) to anticoagulant (Zhang et al., 2008), antitumor (Kaeffer et al., 1998) antihyperlipidemic (Yu et al., 2003) and immunomodulating (Leiro et al., 2007) activities, proved both *in vitro* and *in vivo*.

A brief discussion about the chemical mechanisms that trigger this bioactivity can be worth of mentioning in order to have a deeper insight on the potentiality of using this biomaterial in biomedical applications, and possibly find the "keys" to improve its biological activity.

2.2.1 Antioxidant activity

The research of new antioxidant from renewable natural resources able to scavenge free radicals can represent a virtuous strategy for preventing ROS-induced diseases.

In recent years, several classes of sulphated polysaccharides have been demonstrated to show antioxidant activity. Among them Ulvan extracted from *Ulva pertusa* is reported to play an important role as free radical scavenger *in vitro* and displayed antioxidant activity for the prevention of oxidative damage in living organisms (Qi et al., 2005). As found with other sulphated polysaccharides (Wijesekara et al., 2011) the antioxidant activity is deeply affected by the amount and distribution of sulphate groups inside the Ulvan structure.

The possibility to increase the antioxidant activity of Ulvan can be useful according to the envisaged application and has been successfully investigated both by increasing the degree of sulphation through a sulphur trioxide/N,N-dimethylformamide treatment (Qi et al., 2005) and by introducing suitable groups (acetyl and benzoyl) that can boost the activity of the native polysaccharide (Qi et al., 2006).

2.2.2 Anticoagulant activity

Heparin, a glycosaminoglycan of animal origin containing carboxylic acid and sulphate groups, has been identified and used for more than fifty years as a commercial anticoagulant and it is widely used for the prevention of venous thromboembolic disorders (Pereira et al., 2002). The heparinoid-like structure of Ulvan makes it also able to provide anticoagulant activity. Indeed this class of polysaccharides displayed the inhibition of both the intrinsic pathways of coagulation or thrombin activity and the conversion of fibrinogen to fibrin (Zhang et al., 2008). The molecular weight of the polysaccharide showed an important effect on the anticoagulant activity indicating that longer chains were necessary to achieve thrombin inhibition.

This behavior has been found to be typical of sulphated polysaccharides of marine origins whose anticoagulant activity has been correlated to the content and position of the sulphate groups inside the polymer chains (Melo et al., 2004).

The importance of finding sources of anticoagulants alternative to heparin has been arising due to the associated harmful side effects and the complex steps of purification required to face the immunological concerns and disease transmission associated with its use (Stevens, 2008). Thus the increasing demand for a safer anticoagulant therapy could be potentially satisfied by sulphated polysaccharides obtained from abundant and safer origin and Ulvan can represent the ideal material for such purpose.

2.2.3 Immunomodulating activity

Some of the polysaccharides obtained from the cell walls of seaweeds appear to exert immunomodulatory activities in mammals as they modify the activity of macrophages. Most classes of carrageenans, sulphated polysaccharide obtained from red algae (Figure 1) are known to induce potent macrophages activation (Nacife et al., 2004.).

The structure of the repeating unit mostly found in Ulvan, resembles that typical of glycosaminoglycan, such as hyaluronic acid and chondroitin sulphate because they all contain glucuronic acid and sulphate. Chondroitin sulphate based proteoglycans are known to be produced by macrophages and human monocytes at inflammatory sites (Uhlin-Hansen et al., 1993) and the structure similarity shared with most of sulphated polysaccharides of algal origin and in particular with Ulvan could represent the trigger of the immunomodulating activity of these materials. Indeed Ulvan from *Ulva rigida* has been reported to modulate the activity of murine macrophages and the presence of sulphate groups has demonstrated to be necessary (Leiro et al., 2007). On the other side, a deeper insight into the structure-immunomodulating effect relationship of these polysaccharides would provide the possibility to obtain an anti-inflammatory effect simply by properly modifying their chemical structure.

2.2.4 Antihyperlipidemic activity

Ulvan is known to resist degradation by human endogeneous enzymes thus belonging to the dietary fibers of "sea lattuce" (Taboada et al., 2010). Dietary fibers are considered to be helpful in the prevention of pathologies related to intestinal transit dysfunctions because they act as bulking agents due to their impressive water retention capacity (Bobin-Dubigeon et al., 1997). Dietary fibers are also associated with their ability to lower cholesterol levels (Brown et al., 1999) and the presence of ion charged groups along their structure has shown to improve this beneficial activity (Guillon & Champ, 2000). The ionic groups are thought to complex with bile acids and consequently increase fecal bile acid excretion thus promoting blood cholesterol attenuation (Yu et al., 2003). Ulvan is reported to interact and binding effectively with bile acids due to its high content of negatively charged groups (Lahaye, 1991) thus potentially contributing to the antihyperlipidemic action. Indeed Ulvan has been demonstrated to effectively reduce the level of total and LDL-cholesterol concentrations in the serum and induce an increase in the daily bile excretion in rats (Yu et al., 2003). This activity has been shown to be strongly dependent on the molecular weight of the polysaccharide because a decrease in the viscosity of these materials affected negatively the interaction with bile acids.

3. Potential use of ulvan in biomedical applications

A material suitable for biomedical applications, namely tissue engineering, regenerative medicine and drug delivery, is required to be biocompatible and biodegradable and its products of degradation must be safe and easily cleared from the host organisms. Most of the materials obtained from natural resources are able to fulfil these strict requirements but particular attention are gaining biopolymers of plant and algal origins due to their abundance and minor concerns for purification (Stevens, 2008).

Ulvan could represent an advantageous versatile platform of "unique" sulphated polysaccharides that along with their abundance and renewability would potentially display the properties that match the criteria for biomedical applications. Despite the promising properties related to this material, the use of Ulvan in the biomedical fields is not yet reported and its potentiality still remain unveiled.

A base requirement for a material suitable for tissue engineering, regenerative medicine and/or drug delivery is its insolubility in the physiological fluids not possible with most classes of polysaccharides, whose high hydrophilicity make them very akin to water molecules. Indeed materials used for the regeneration of organs or tissues must avoid dissolution in contact with body fluids thus functioning as chemically and mechanically stable scaffolds during the growth and differentiation of the implanted cells. Also polymeric materials used for drug delivery must preserve their integrity or degrade slowly in order to maintain a controlled release of the loaded drug. The use of polysaccharides in these types of applications is possible only after proper chemical modifications aimed at making them insoluble in aqueous solution. A possible strategy consists in decreasing the hydrophilicity of these materials by introducing hydrophobic groups in their structures, but this often leads to the obtainment of new class of materials, mostly semi-synthetic than naturals and often very different from the original biomaterials. The strategy of election mostly followed by biomaterial scientists in the last 50 years (Hoffmann, 2002) consisted simply in the induction of "junction-zones" between the polymeric chains, inhibiting their dissolution through the formation of permanent or temporary crosslinked networks displaying hydrogel features. These structures maintain almost completely the chemical properties of the original biopolymer comprising their affinity to water giving rise to swollen and not dissolved polymeric scaffolds of natural origins. The maintained hydrophilic character of hydrogels is particularly important in tissue engineering where the overall permeation of nutrients and cellular products into the pores of the gel, determinant for the growth and differentiation of the cells, is determined by the amount of water in the structure (Hoffmann, 2002).

Hydrogels can be chemically or covalently crosslinked and are defined as permanent when the "junction-zones" between the constituting chemical chains are formed by covalent links. If the new bonds are not susceptible to hydrolysis or enzyme recognition the formed hydrogel can be stable indefinitely and not prone to degradation.

The formation of physical or temporary hydrogels is triggered when the "junction-zones" between the polymeric chains are stabilized by weak forces such as electrostatic or hydrophobic interactions. These interactions are reversible, and can be disrupted by changes in physical conditions such as ionic strength, pH, temperature or application of stress. Physical hydrogels are usually not homogeneous, since clusters of molecular entanglements or hydrophobically- or ionically-associated domains can create in-homogeneities (Hoffmann, 2002). This can lead to the formation of hydrogels with weak mechanical properties not suitable for most conventional applications.

Apart from the biological activity displayed by these biomaterials, other important attributes of hydrogels are their mechanical properties and degradation rates that must be tuned according to the final application. The degree of crosslinking along with the chemical nature of the polymer and of the „junction-zones" represent the key parameters that mainly affect the above-mentioned properties and can be adjusted according to the addressed applications.

Hydrogels based on polysaccharides are usually characterized by poor mechanical properties due to their impressive water uptake and swelling that lead to the formation of wide opened pore structures that ultimately weakens the scaffold architecture (LaNasa et al., 2010). The presence of charged groups like sulphate and carboxylate in the structure of Ulvan would lead to an even more accentuated absorption of water molecules, hampering the preparation of mechanically stable hydrogels. The strategies for increasing the mechanical properties are several and comprise the preparation of interpenetrating networks with other polymers like polycomplexes formed between oppositely charged polyelectrolytes (Hamman, 2010), the preparation of composite hydrogels mixed with inorganic additives (Pavlyuchenko & Ivanchev, 2009) or the use of hydrophobic comonomers as in the preparation of hydrogels by radical crosslinking (Li et al., 2003).

All these strategies strongly affect the chemical nature of the original biopolymers and the properties of the final hydrogels can be very different from those of the native materials comprising their biocompatibility and bioactivity. A strategy to improve the mechanical properties of polysaccharides and in particular Ulvan is based on a proper choice of the nature and amount of "junction-zones" that crosslink the polymer. Indeed both the increase of the crosslinking or the preference for chemical instead of physical crosslinking would positively affect the mechanical properties of the final hydrogels, leading to more compact structures and dimensionally shaped architecture. Both type of crosslinking have been conducted on Ulvan and are worth of mentioning to get a deeper insight on the possible applications and future developments of using this biopolymer in the biomedical fields.

3.1 Ulvan-based physical hydrogels

Ionotropic gelation is a kind of physical crosslinking based on the ability of polyelectrolytes to give hydrogels in the presence of counter-ions. Alginate is a naturally occurring polysaccharide obtained from marine brown algae that spontaneously form reticulated structures in the presence of divalent or polyvalent cations (Patil et al., 2010). The mechanism involves the cooperative interactions of the carboxylate groups of alginate with the polyvalent cations present in solution to form "junction-zones" between the chains that crosslink the matrix in an insoluble polymeric network.

Due to its polyanionic nature Ulvan is expected to show a similar behavior but its gel formation has the unique characteristic of involving borate esters (Haug, 1976). The optimal conditions for the preparations of the hydrogels requires the presence of boric acid and calcium ions at slightly basic conditions (pH 7.5) giving hydrogels with storage modulus of about 250 Pa (Lahaye & Robic, 2007). Higher ion concentrations, different pH, and even phosphate buffering ions are detrimental to the gel.

The mechanism of gel formation is not yet completely unveiled but is proposed to proceed through the formation of borate esters with Ulvan 1,2-diols followed by crosslinking via Ca^{2+} ions (Haug, 1976). Since the gel is thermoreversible the "junction-zones" that crosslink the polymer are thought to involve weak linkages, likely based on labile borate ester groups and ionic interactions easily disrupted by thermal treatments (Lahaye & Robic, 2007). Calcium ions would bridge complexes and/or stabilize the borate esters (Fig. 4a) but also sulphate and carboxylic acid groups were later on proposed to coordinate to Ca^{2+} (Figure 3b,c) (Lahaye & Axelos, 1993) and contribute to the gel formation.

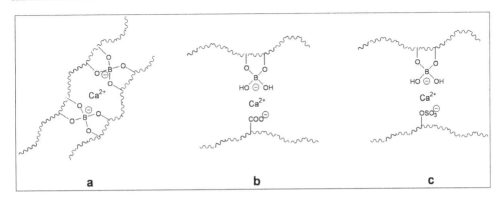

Fig. 4. Proposed mechanisms for the formation of Ulvan-based ionotropic hydrogels through a) Ca²⁺ stabilization of borate esters (Haug, 1976) or the participation of b) carboxylate and c) sulphate groups (Lahaye & Axelos, 1993).

The gel behaviour of Ulvan is different from that of most polysaccharides which usually involve tight junction zones of ordered molecular structures like helices or flat buckled ribbons (Stephen, 1995). Ulvan gel results from the aggregation of bead-like structures interconnected by more hydrophilic polymeric fractions. This behaviour would be undoubtedly favoured by the necklace-type ultrastructure assumed by the polymer in solution, whose formation has been shown to be promoted by ionic interactions (Robic et al, 2009). To this view, the positive role of boric acid can also be related to its reactivity towards the neutrally charged hydroxyl moieties of Ulvan and subsequent substitution with charged borate ester groups, thus contributing to create additional charges on the beads surface of the polysaccharide and favouring their association.

The mechanical properties of these ionotropic hydrogels are usually poor due to their intrinsic weakness and tend to get worse when used in contact with body fluids due to the ion exchange phenomena that occurs between Ca^{2+} that stabilizes the network and the monovalent cations like K^+ and Na^+ present in the physiological liquids (LeRoux et al., 1999). These hydrogels found limited applications in tissue engineering due to their mechanical instability and uncontrolled dissolution in physiological conditions (Atala & Lanza, 2002).

3.2 Ulvan-based chemical hydrogels

In order to overcome the problems related to the mechanical instability of the physically gelled Ulvan and extend the range of their potential applications, the strategy of chemical crosslinking of Ulvan was undertaken (Morelli & Chiellini, 2010).

A smart and relatively innovative technique of obtaining chemically crosslinked hydrogels is represented by their photopolymerization in the presence of photoinitiators using visible or ultraviolet (UV) light.

Photopolymerization is used to convert a liquid monomer or macromer to a hydrogel by free radical polymerization in a fast and controllable manner under ambient or physiological conditions. The mechanism is triggered by visible or UV light that interact with light-sensitive compounds called photoinitiators to create free radicals that can initiate polymerization of species containing suitably reactive groups (typically double bonds).

Photopolymerization has several advantages over conventional polymerization and crosslinking techniques. These include spatial and temporal control over polymerization, fast curing rates at room or physiological temperatures and minimal heat production (Nguyen & West, 2002). Moreover photopolymerization does not require the use of many reactive species, initiator and catalysts usually involved in the conventional chemical crosslinking methods, thus representing a potentially safer technique.

The condition for UV photopolymerization is that polymeric materials need to be conjugated with radically polymerizable groups. In this context, methacryloyl or acryloyl groups, when grafted to the chain backbone via an oxygen or a nitrogen atom usually represent a good candidates for this function, as they work as degradable crosslinks sensitive to either hydrolysis (Benoit et al., 2006) or cell-mediated proteolysis (Mahoney & Anseth, 2006).

The introduction of vinyl or vinylidenic polymerizable groups on Ulvan has been conducted by using several different (meth)acryloyl precursors and conditions (Morelli & Chiellini, 2010) (Figure 5).

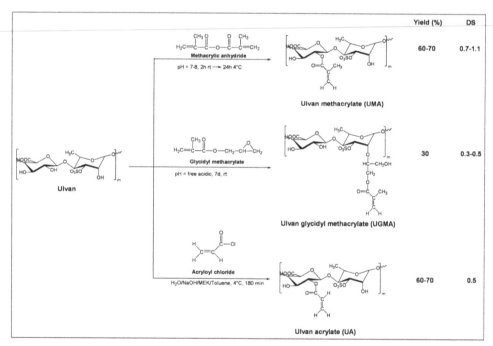

Fig. 5. Reaction of Ulvan with (Meth)acryloyl precursors and relative experimental conditions. Next to every macromer preparation are reported the mean values of the yield (%) of the final products and the degree of substitution (DS) expressed as the mean number of (meth)acryloyl group present in every repeating unit.

The reaction of Ulvan with organic chemical precursors is not straightforward because partially hampered by its insolubility in the common organic solvents. This compels its modification under heterogeneous and not favourable conditions as the ones reported in Figure 5. Also the aggregative behaviour of Ulvan in aqueous solutions limits its "reactive surface" available for modifications and this may be the cause for the low degree of (meth)acryloyl substitution usually found on the final macromers.

The amount of UV polymerizable unsaturated groups introduced onto the polysaccharide would definitively affect the physical properties of the final hydrogels because it determines the number of "junction-zones" that crosslink the linear polymer chains. Both the substitution of the polar hydroxyl groups with the hydrophobic (meth)acryloyl moieties and the increase of the number of crosslinks inside the hydrogel structures would determine a minor absorption of water molecules with a consequent improvement in the mechanical properties of the hydrogels (Anseth et al., 1996). To this view the mean number of crosslinkable groups present in every repeating unit of Ulvan – expressed as substitution degree (SD) - represents a key parameter to be evaluated.

The calculation of the substitution degree (SD) of the macromers obtained by the different chemical routes reported in Figure 5, has been performed by ^{1}HNMR analyses (Figure. 6).

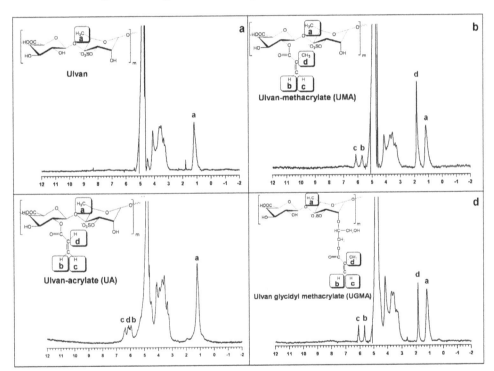

Fig. 6. ^{1}HNMR spectra in D$_2$O of : a) Ulvan, b) Ulvan-methacrylate (UMA), c) Ulvan-acrylate (UA). d) Ulvan-glycidyl methacrylate (UGMA). Chemical structures of the main disaccharide repeating units of the polysaccharides are reported together with the relative peak assignment as highlighted in the small boxes.

The SD has been calculated by comparing the peak areas relative to the vinyl protons of the introduced (meth)acryloyl groups (Fig. 6b-d) with the peak area relative to the methyl group of the rhamnose present in the native Ulvan (Fig. 6a). The SD values reported in Figure 5 represent only a rough estimation of the actual values because the chemical structure of the repeating unit of Ulvan used for the calculations (Fig. 6a) can not be considered univocal (Lahaye & Robic, 2007).

The most effective procedure for the preparation of macromers has been shown to utilize a large excess of methacrylic anhydride in slightly basic conditions for 24 hours at 4°C (Fig. 5). A fairly large amount of reactive is required because of the competitive hydrolysis that spoils the methacryloyl precursor during the reaction. The formation of the methacryloyl derivative of Ulvan was further confirmed by FT-IR analysis (Fig. 7).

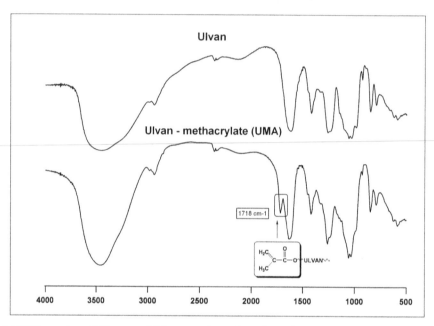

Fig. 7. FT-IR spectra of Ulvan and Ulvan methacrylate (UMA) macromer with the structure of the conjugated methacryloyl group and the relative absorption frequency reported in small boxes.

The spectra of the native Ulvan and the Ulvan methacrylate derivative are completely overlapping except for the peak at 1718 cm-1 likely attributable to the presence of an α,β-unsaturated carboxylic ester.

The preparation of hydrogels has been carried out by using a small amount of a cytocompatible photoinitiator – IRGACURE ® 2959 – and exposing the polymeric solution to UV light – 365 nm – at short irradiation times. The mechanism of the covalent crosslinking between the polymeric chains involves the radical polymerization of the conjugated (meth)acryloyl groups and the formation of degradable carboxylic ester based "junction-zones" that act as crosslink moieties (Figure 8).

The crosslinking degree (CD) of Ulvan macromers have been evaluated by ^{1}HNMR analyses of the solutions before and after definite times of irradiation by monitoring the peak areas of the reacting unsaturated protons (**b** and **c**, Figure 9). This technique has been applied to UMA because the preparation of this macromer proved to be the most effective in terms of yield and SD.

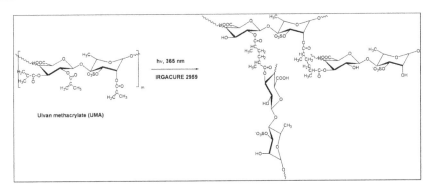

Fig. 8. Photopolymerization of Ulvan methacrylate under UV irradiation – 365 nm – in presence of a photoinitiator.

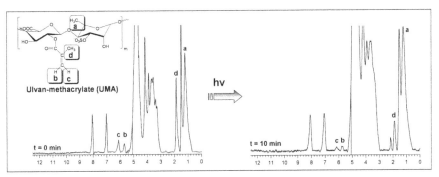

Fig. 9. ¹HNMR spectra of Ulvan-methacrylate (UMA) before and after 10 min of UV irradiation in the presence of IRGACURE® 2959 as photoinitiator. Chemical structures of the main disaccharide repeating units of the macromer is reported together with the relative peak assignment as highlighted in the small boxe.

The exposure of Ulvan macromers to UV light Leads definitely to the formation of hydrogels but the photopolymerization is not complete even after 10 minutes of irradiation as demonstrated by the residual peak of the unreacted double bond protons **c** and **b** in the ¹HNMR spectrum. The degree of crosslinking has been monitored at fixed irradiation times and showed that half (meth)acryloyl groups are polymerized after 5 minutes of UV curing (Table 1).

UV exposure time (min)	CD (%)
1	22.5
2	31.5
3	38.0
5	51.0
10	68.5

Table 1. Crosslinking degree (CD) for UMA having SD = 1 under UV irradiation (365 nm, 8 mW · cm⁻²) as a function of exposure time.

The uncomplete crosslinking of Ulvan macromers under UV exposure could be ascribed to both the nature and the morphology assumed by this polysaccharide in solution. Indeed the antioxidant activity of Ulvan could reduce the rate of radical polymerization of the macromers by quenching the radicals formed during the UV irradiation. Moreover the aggregative behaviour of Ulvan in aqueous solution reduces the amount of (meth)acryloyl groups available for polymerization thus partially inhibiting the crosslinking.

The property of retaining water represents a key parameter for evaluating the quality of a hydrogel and its potential use for biomedical applications because it usually affects its permeability, biocompatibility and rate of degradation. The swelling ability of hydrogels could also provide information about their mechanical stability and chemical and physical properties, since the degree of water uptake is related both to the chemical nature and to the physical structure of the polymeric network (Qi et al., 2004). It is known, for example, that gels exhibiting a larger pore structure – likely due to a lower degree of crosslinking – have poor mechanical strength and higher swelling ratios (Anseth et al., 1996).

The swelling ability of hydrogels is usually quantified by measuring their Swelling Degree % (SD%) taken as the ratio (%) between the weight of the swollen hydrogel to that of the dried sample. The swelling degrees of the prepared Ulvan hydrogels have been carried out in phosphate buffer solutions (pH 7.4) and their behaviour was recorded during 7 days of immersion (Fig. 10).

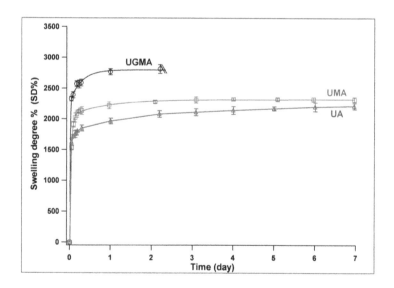

Fig. 10. SD% in PBS buffer solution (0.1 M, pH 7.4) of UV crosslinked (365 nm, approximately 8 mW · cm⁻²) UMA and UGMA hydrogels as a function of time.

Pictures of the swollen scaffolds taken after 2 days of immersion in phosphate buffer saline (PBS) at pH 7.4, showed that the Ulvan methacrylate (UMA) hydrogels proved to be most stable in terms of texture and mechanical properties (Figure. 11).

The swelling degree experiments of UGMA-based samples were stopped after 2 days of immersion since the hydrogels were no longer coherent and hence not easy to handle. This behaviour could be interpreted by considering the low degree of substitution typically obtained with this type of macromer (Figure 5) and subsequent low degree of crosslinking. The final hydrogels resulted more hydrophilic, as demonstrated by the highest swelling degree values (SD%) obtained, and less crosslinked thus leading to network with weak mechanical properties.

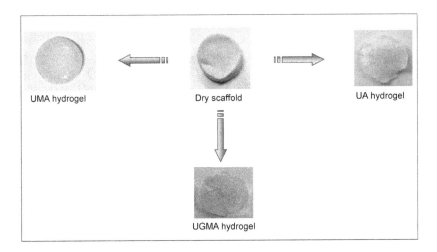

Fig. 11. Pictures of the swollen hydrogels obtained after 2 days of immersion in the swelling medium (phosphate buffer, 0.1M, pH 7.4).

The swelling degree values obtained with UMA- and UA-based macromers resulted to be similar thus indicating a similar density of crosslinking inside their structure. The interpretation of these datas are not straightforward since UMA-based hydrogels were expected to be less hydrophilic and more crosslinked due to the higher amount of polymerizable group contained in their structure. Indeed the final texture of the UMA-based hydrogels indicated a better mechanical stability in respect to the other type of hydrogels. The unexpected lower SD (%) values obtained with UA-based hydrogels could be explained by the loss of material during the swelling experiments.

4. Conclusion

Ulvan, a sulphated polysaccharide of algal origin, is worth of deeper attention for its potential use in technological and industrial-related applications. The exploitation of this abundant and renewable resource could represent an advantageous alternative approach to the use of fully synthetic materials based on fossil fuel feedstock.

In particular Ulvan represents an intriguing candidate material for biomedical applications due to its intrinsic beneficial biological activities and the possibility of easily modifying its structure according to the envisaged application. Its chemical structure similar to that of natural glycosaminoglycans such as chondroitin sulphate and hyaluronic acid make Ulvan an attractive candidate for their substitution or use in related applications.

In order to be employed in biomedical applications such as tissue engineering, regenerative medicine and drug delivery, Ulvan needs to be converted into an insoluble material under physiological conditions and to have mechanical properties suitable for the end application. The preparation of physically crosslinked Ulvan hydrogels has been reported since long times, but their weak mechanical properties and uncontrolled dissolution in presence of physiological fluids make them unsuitable for biomedical uses, where a scaffolding role is required

A novel method for covalent crosslinking of Ulvan through the UV mediated radical polymerization of activated macromers by double bond conjugated moieties, revealed to be promising in the preparation of chemically crosslinked Ulvan hydrogels. The conjugation of methacryloyl group to Ulvan through the reaction with methacrylic anhydride under slightly basic conditions gave the best results in terms of product yield and substitution degree. The hydrogels obtained after their exposure to UV light seemed to be very stable in physiological conditions.

The crosslinking of the Ulvan macromer precursors is usually not complete because is hampered both by its aggregative behaviour in solution that limits the availability of the (meth)acryloyl groups and very presumably by the radical quenching activity of the polysaccharide during the UV exposure thus negatively affecting the mechanical properties of the final hydrogels. Nevertheless the antioxidant activity of Ulvan could make this material a good candidate as a matrix for cell encapsulation due to the possible protection against the radicals produced during UV crosslinking (Fedorovich et al., 2009). The use of these materials as a base for cytocompatible scaffolds is also promoted by the softness related to partial crosslinking of these macromers, since it is known that cell spreading within hydrogels is influenced by matrix stiffness and soft matrices interestingly are expected to promote cell spreading (Liu & Chan-Park, 2009).

Moreover the possibility of preparing Ulvan based hydrogels by a straightforward technique such as UV crosslinking makes the use of Ulvan in biomedical fields even more attractive. Indeed UV photopolymerization allows the spatial and temporal control over the crosslinking and the fabrication of hydrogels in situ with the possibility of forming complex architectures that adhere and conform to tissue structure.

5. Acknowledgment

The present work was performed within the framework of NOE project Expertissues (NMP3-CT 2004-500328) and Expertissues Miniproject NATCOM.

6. References

Alban, S., Schauerte, A., Franz, G. (2002). Anticoagulant sulphated polysaccharides: Part I. Synthesis and structure–activity relationships of new pullulan sulphates. *Carbohydrate Polymers*, Vol.47, No.3, (February 2002), pp. 267-276, ISSN 0144-8617.

Alsberg, E., Anderson, K.W., Albeiruti, A., Franceschi, R.T., Mooney, D.J. (2001). Cell-interactive Alginate Hydrogels for Bone Tissue Engineering *Journal of Dental Research*, Vol.80, No.11, (November 2001), pp. 2025-2029, ISSN 1462-6446.

Alves, A., Caridade, S.G., Mano, J.F., Sousa, R.A., Reis, R.L. (2010). Extraction and physico-chemical characterization of a versatile biodegradable polysaccharide obtained from green algae, *Carbohydrate Research*, Vol.345, No.15, (October 2010), pp. 2194-2200, ISSN 0008-6215.

Anseth, K.S, Bowman, C.N., Peppas, L.B. (1996). Mechanical properties of hydrogels and their experimental determination *Biomaterials*, Vol.17, No.17, (January 1996), pp. 1647-1657, ISSN 0142-9612.

Athukorala, Y., Ahn, G.N., Jee, Y.H., Kim, G.Y., Kim, S.H., Ha, J.H., (2008) Antiproliferative activity of sulphated polysaccharide isolated from an enzymatic digest of Ecklonia cava on the U-937 cell line, *Journal of Applied Phycology*, Vol.21, No. 3, (August 2008), pp. 307-314, ISSN 0273-2289.

Atala, A., Lanza, R.P. (2002). *Methods of Tissue Engineering*, Academic Press, ISBN 978012436636, San Diego, CA.

Barghini, A., Ilieva, V.I., Imam, S.H., Chiellini , E. (2010) PCL and PHB blends containing seaweed fibers: Morphology and Thermal-Mechanichal Properties", *Journal of Polymer Science Part A: Polymer Chemistry*, Vol. 48, No. 23, (December 2010), pp. 5282-5288, ISSN 1520 - 5738.

Benoit, D.S, Durney, A.R., Anseth, K.S. (2006). Manipulations in Hydrogel Degradation Behavior Enhance Osteoblast Function and Mineralized Tissue Formation, *Tissue Engineering*, Vol. 12, No. 6, (June 2006), pp. 1663-1673. ISSN 1076-3279.

Bobin-Dudigeon, C., Lahaye, M., Guillon, F., Barry, J.L., Gallant, D. (1997). Factors limiting the biodegradation of *Ulva* sp cell-wall polysaccharides *Journal of the Science of Food and Agriculture*, Vol.75, No.3, (November 1997), pp. 341-351, ISSN 1097-0010.

Bouhadir, K.H., Lee, K.Y, Alsberg, E., Damm, K.L., Anderson, K.W., Mooney, D.J. (2001). Degradation of Partially Oxidized Alginate and Its Potential Application for Tissue Engineering *Biotechnology Progress*, Vol.17, No.5, (September 2001), pp. 945-950, ISSN 1520-6033.

Brand, X., Morand, P., (1997) Anaerobic digestion of *Ulva* sp. 1. Relationship between *Ulva* composition and methanisation, *Journal of Applied Phycology*, Vol.9, No. 6, (December 1997), pp. 511-524, ISSN 0273-2289.

Brown, L., Rosner, B., Willet, W.W, Sachs, F.M. (1999). Cholesterol lowering effect of dietary fiber: a meta-analysis, *American Journal of Clinical Nutrition*, Vol. 69, No. 1, (January 1999), pp. 30-42 ISSN 0002-9165.

Chiellini, E., Cinelli, P., Ilieva, V.I., Martera, M. (2008) Biodegradable Termoplastic Composites Based on Polyvinyl Alcohol and Algae, *Biomacromolecules*, Vol. 9, No. 3, (March 2008) pp. 1007 – 1013, ISSN 1525-1797.

Chiellini, E., Cinelli, P., Ivanova Ilieva, V., Zimbardi, F., Kanellopoulos, N., de Wilde, B., Pipino, A., Anders, B., (2009) Hybrid Composites Based on Fibres of Marine Origin, Int. J. Materials and Product Technology, Vol. 36, No. 1-2-3-4, (January 2009), pp. 47-61 ISSN 0268-1900

Christi, Y., (2007). Biodiesel from microalgae *Biotechnology Advances*, Vol.25, No.3, (February 2007), pp. 294-306, ISSN 0734-9750.

Chung, T.W., Yang, J., Alkaike, T., Cho, K.Y., Nah, J.W., Kim, S.I., Cho, C.S. (2002). Preparation of alginate/galactosylated chitosan scaffold for hepatocyte attachment *Biomaterials*, Vol. 23, No.14, (July 2002), pp. 2827-2834, ISSN 0142-9612.

Damonte, E.B., Matulewicz, M.C., Cerezo, A.S. (2004). Sulphated seaweed polysaccharides as antiviral agents *Current Medicinal Chemistry*, Vol.11, No.18, (September 2004), pp. 2399-2419, ISSN 0929-8673.

Dar, A., Shachar, M., Leor, J., Cohen, S. (2002). Optimization of cardiac cell seeding and distribution in 3D porous alginate scaffolds *Biotechnology and Bioengineering*, Vol.80, No.3, (November 2002), pp. 305-312, ISSN 1097-0290.

D'Ayala, G., Malinconico, M., Laurienzo, P., (2008). Biodiesel from microalgae *Molecules*, Vol.13, No.9, (September 2008), pp. 2069-2106, ISSN 1420-3049.

Dobrynin, A.V. (2008). Theory and simulations of charged polymers: From solution properties to polymeric nanomaterials *Current Opinion in Colloid & Interface Sciences*, Vol.28, No.6, (December 2008), pp. 1859-1871, ISSN 0024-9297.

Dobrynin, A.V., (1995). Scaling theory of polyelectrolyte solutions *Macromolecules*, Vol.13, No.6, (March 1995), pp. 376-388, ISSN 1359-0294.

Demirbas, A., Demirbas, M.F. (2011). Importance of algae oil as a source of biodiesel *Energy Conversion and Management*, Vol.52, No.1, (January 2011), pp. 163-170, ISSN 0196-8904.

Fedorovich, N.E.., Oudshoorn, M.H., van Geemen, D., Hennink, W.E., Wouter, J.A., Dhert, J.A. (2009). The effect of photopolymerization on stem cells embedded in hydrogels, *Biomaterials*, Vol. 30, No. 3, (January 2009), 344-353, ISSN 0142-9612.

Fletcher, R.L., (1996). *Marine Benthic Vegetation : Recent Changes and the Effect of Eutrophication*, Springer, ISBN 978-3-540-58106-2, Berlin, Germany.

Guillon, F., Champ, M. (2000). Structural and physical properties of dietary fibres, and consequences of processing on human physiology. *Food Research International*, Vol. 33, No. 3-4, (April 2000), pp. 233-245, ISSN 0963-9969.

Hamman, J.S. (2010). Chitosan Based Polyelectrolyte Complexes as Potential Carrier Materials in Drug Delivery Systems. *Marine Drugs*, Vol. 8, No. 4, (March 2010), pp. 1305-1322, ISSN 1660-3397.

Hashimoto, T., Suzuki, Y., Yamamoto, E., Tanihara, M., Kakimaru, Y., Suzuki, K. (2004). Development of alginate wound dressings linked with hybrid peptides derived from laminin and elastin *Biomaterials*, Vol.25, No.7-8, (March 2004), pp. 1407-1414, ISSN 0142-9612.

Hoffmann, A.S. (2002). Hydrogels for biomedical applications. *Advanced Drug Delivery Reviews*, Vol. 54, No. 1, (January 2002), pp. 3-12, ISSN 0169-409X.

Haug, A. (1976) The influence of borate and calcium on the gel formation of a sulfated polysaccharide from Ulva lactusa, *Acta Chemica Scandinavica*, Vol. B30, No. 6 (January 1976), pp. 562–566, ISSN 0001-5393.

John, R.P., Anisha, J.S., Madhavan Nampoothiri, K., Pandey, A. (2011). Micro and Macroalgal biomass: A renewable source for bioethanol *Bioresource Technology*, Vol.102, No.1, (January 2011), pp. 186-193, ISSN 0960-8524.

Kaeffer, B., Benard, C., Lahaye, M., Blottiere, H.M., Cherbut, C., Biological properties of ulvan, a new source of green seaweed sulphated polysaccharides, on cultured normal and cancerous colonic epithelial cells, *Planta Medica*, Vol. 65, No. 6, (August 1999), pp. 527-531, ISSN 0032-0943.

Lahaye, M. (1991). Marine algae as sources of fibres: Determination of soluble and insoluble dietary fibre contents in some 'sea vegetables, *Science of Food and Agriculture*, Vol. 54, No. 4, (Janauary 1991), pp. 587-594, ISSN 0022-5142.

Lahaye, M., Axelos, M.A.V. (1993). Gelling properties of water-soluble polysaccharides from proliferating marine green seaweeds (Ulva spp.), *Carbohydrate Polymers*, Vol.22, No. 4, (February 1993), pp. 261-265, ISSN 0144-8617.

Lahaye, M., Brunel, M., Bonnin, M. (1997). Fine chemical structure analysis of oligosaccharides produced by an ulvan-lyase degradation of the water-soluble cell-wall polysaccharides from *Ulva* sp. (Ulvales, Chlorophyta) *Carbohydrate Polymers*, Vol.304, No.3-4, (November 1997), pp. 261-265, ISSN 0144-8617.

Lahaye, M., Jegou, D., Buleon, A., (1994). Chemical characteristics of insoluble glucans from the cell wall of the marine green alga Ulva lactuca (L.) Thuret , *Carbohydrate Research*, Vol.262, No.1, (September 1994), pp. 115-125, ISSN 0008-6215

Lahaye, M., Robic, A. (2007). Structure and functional properties of ulvan, a polysaccharide from green seaweeds, *Biomacromolecules*, Vol.8, No.6, (April 2007), pp. 1765-1774, ISSN 1525-7797.

LaNasa, S.M., Hoffecker, I.T., Briant, S.J. (2010). Presence of pores and hydrogel composition influence tensile properties of scaffolds fabricated from well-defined sphere templates, *Journal of Biomedical Materials Research Part B: Applied Biomaterials*, Vol. 96B, No. 2, (July 2010), pp. 294-302, ISSN 1552-4981.

Leiro, J.M., Castro, R., Arranz, J.A., Lamas, J. (2007). Immunomodulating activities of acidic sulphatated polysaccharides obtained from the seaweed Ulva rigida C. Agardh, *International Immunopharmacology*, Vol.7, No.7, (July 2007), pp. 879-888, ISSN: 1567-5769.

LeRoux, M.A., Guilak, F., Setton, L.A. (1999) Compressive and shear properties of alginate gels: Effect of sodium ions and alginate concentration, *Journal of Biomedical Materials Research*, Vol. 47, No. 1, (October 1999), pp. 46-52, ISSN 1552-4965

Li, Q., Wang, D., Helisseeff, J.H. (2003). Heterogeneous-phase reaction of glycidyl methacrylate and chondroitin sulphate, *Macromolecules*, Vol. 36, No. 7, (April 2003), pp. 2556-2562, ISSN 0024-9297.

Liu, Y., Chan-Park, M.B. (2009). Hydrogel based on interpenetrating polymer networks of dextran and gelatin for vascular tissue engineering, *Biomaterials*, Vol. 30, No. 2, (January 2009), 196 - 207, ISSN 0142-9612.

Mahoney, M.J., Anseth, K.S. (2006). Three-dimensional growth and function of neural tissue in degradable polyethylene glycol hydrogels, *Biomaterials*, Vol. 27, No. 10, (April 2006), 2265-2274, ISSN 0142-9612.

Mano, J.F., Silva, G.A., Azevedo, H.S., Malafaya, P.B., Sousa, R.A., Silva, S.S., Boesel, L.F., Oliveira, J.M., Santos, T.C., Marques, A.P., Neves, N.M., Reis, R.L. (2007). Natural origin biodegradable systems in tissue engineering and regenerative medicine: present status and some moving trends *Journal of the Royal Society Interface*, Vol. 22, No. 4, (December 2007), pp. 999-103o, ISSN 1742-5662.

Matsumoto, M., Yochouchi, H., Suzuki, N., Ohata, H., Matsunaga, T. (2003). Saccharification of marine microalgae using marine bacteria for ethanol production *Journal of Applied Phycology*, Vol. 105, No.1-3, (February 2003), pp. 247-254, ISSN 0273-2289.

Mazè, J., Morand, P., Potoky, P. (1993). Stabilisation of 'green tides'Ulva by a method of composting with a view to pollution limitation. *Journal of Applied Phycology*, Vol. 5, No. 2, (April 1993), pp. 183-190, ISSN 0273-2289.

Melo, F.R., Pereira, M.S., Foguel, D., Mourao, P.A.S. (2004) Antithrombin-mediated anticoagulant activity of sulphated polysaccharides, *Journal of Biological Chemistry*, Vol. 279, No. 20 (May 2004), pp. 20824–20835, ISSN: 0021-9258.

Morand, P., Briand, X. (1996). Excessive Growth of Macroalgae: A Symptom of Environmental Disturbance, *Botanica Marina*, Vol. 39, No. 1-6, (January 1996), pp. 491-516, ISSN 1437-4323.

Morand, P., Briand, X., Charlier, R.H. (2006). Anaerobic digestion of Ulva sp. 3. Liquefaction juice extraction by pressing and a technico-economic budget *Journal of Applied Phycology*, Vol.18, No.6, (February 2006), pp. 741-755, ISSN 0921-8971.

Morelli, A., Chiellini, F. (2010). Ulvan as a New Type of Biomaterial from Renewable Resources: Functionalization and Hydrogel Preparation, *Macromolecular Chemistry and Physics*, Vol. 211, No. 7, (April 2010), pp. 821-832, ISSN 1022-1352.

Nguyen, K.T., West, J.L. (2002). Photopolymerizable hydrogels for tissue engineering applications, *Biomaterials*, Vol. 23, No. 22, (November 2002), 4307-4314, ISSN 0142-9612.

Nacife, V.P., Soeiro, M.D., Gomes, R.N., D'Avila, H., Castro-Faria Neto, H.C., Meirelles, M.N.L. (2004) Morphological and biochemical characterization of macrophages activated by carrageenan and lipopolysaccharide in vivo. *Cell Structure and Function*, Vol. 29, No. 2 (September 2004), pp. 27–34, ISSN 0386-1796.

Paradossi, G., Cavalieri, F., Chiessi, E. (2002). A conformational study on the algal polysaccharide ulvan *Macromolecules*, Vol.35, No.16, (July 2002), pp. 6404-6411, ISSN 0024-9297.

Paradossi, G., Cavalieri, F., Pizzoferrato, L., Liquori, A.M., (1999). A physico-chemical study on the polysaccharide ulvan from hot water extraction of the macroalga Ulva, *International Journal of Biological Macromolecules*, Vol.25, No.4, (August 1999), pp. 309-315, ISSN 0141-8130.

Patil, J.S., Kamalapur, M.V., Marapur, S.C., Kadam, D.V. (2010). Ionotropic gelation and polyelectrolyte complexation: the novel technique to design hydrogel particulate sustained, modulated drug delivery system: a review. *Digest Journal of Nanomaterials and Biostructures*, Vol. 5, No. 1, (March 2010), pp. 241-248, ISSN 1842-3582.

Pavlyuchenko, V.N., Ivanchev, S.S., (2009). Composite polymer Hydrogels, *Polymer Science Series A*, Vol. 51, No. 7, (July 2009), pp. 743-760, ISSN 0965-545X.

Pereira, M.S., Melo, F.R., Mourao, P.A.S., (2002). Is there a correlation between structure and anticoagulant action of sulfated galactans and sulfated fucans?, *Glycobiology*, Vol. 12, No. 10, (October 2002), pp. 573-580, ISSN 0959-6658.

Qi, H., Zhang, Q., Zhao, T., Chen, R., Zhang, H., Niu, X. Antioxidant activity of different sulphate content derivatives of polysaccharide extracted from Ulva pertusa (Chlorophyta) in vitro, (2005), *International Journal of Biological Macromolecules,* Vol.37, No.4, (December 2005), pp. 195-199, ISSN 0141-8130.

Qi, H., Zhang, Q., Zhao, T., Hu, R., Zhang, K., Li, Z., In vitro antioxidant activity of acetylated and benzoylated derivatives of polysaccharide extracted from Ulva pertusa (Chlorophyta), (2006), *Bioorganic Medicinal Chemistry Letters,* Vol.16, No.9, (May 2006), pp. 2441-2445, ISSN 0960-894X.

Qi, L., Williams, C.G., Sun, D.D.N., Wang, J., Leong, K., Elisseeff, J.H. (2004) *Journal of Biomedical Material Research: Part A,* Vol. 68A, No. 1, (January 2004), pp. 28-33, ISSN 1552 -4965.

Ricciardi, R., Auriemma, F., De Rosa, C. (2005). Structure and Properties of Poly(vinyl alcohol) Hydrogels Obtained by Freeze/Thaw Techniques *Macromolecular Symposia,* Vol.222, No.1, (March 2005), pp. 49-64, ISSN 1521-3900.

Rioux, L.E., Turgeon, S.L., Beaulieu, M. (2007). Characterization of polysaccharides extracted from brown seaweeds *Carbohydrate Polymers,* Vol.69, No.3, (June 2007), pp. 530-537, ISSN 0144-8617.

Robic, A., Gaillard, C., Sassi, J.F., Lerat, Y., Lahaye, M. (2009). Ultrastructure of Ulvan: A polysaccharide from green seaweeds *Biopolymers,* Vol.91, No.8, (August 2009), pp. 652-664, ISSN 0006-3525.

Rocha de Souza, M.C., Marques, C.T., Dore, C.M.G., Ferreira da Silva, F.R., Rocha, H.A.O, Leite, E.L., (2007). Antioxidant activities of sulphated polysaccharides from brown and red seaweeds *Journal of Applied Phycology,* Vol.19, No.2, (April 2007), pp. 153-160, ISSN 0921-8971.

Stephen, A.M., (1995). *Food Polysaccharides and their applications,* Marcel Dekker Inc., ISBN 0824793536, New York, USA.

Stevens, M.M. (2008). Biomaterials for bone tissue engineering, *Materials Today,* Vol.11, No.5, (May 2008), pp. 18-25, ISSN 1369-7021

Taboada, C., Millan, R., Miguez, I. (2010). Composition, nutritional aspects and effect on serum parameters of marine algae Ulva rigida, *Journal of the Science of Food and Agriculture,* Vol.90, No.3, (February 2010), pp. 445-449, ISSN 1097-0010.

Tsai, C.S., (2007). *Biomacromolecules : introduction to structure, function, and informatics,* John Wiley & Sons, ISBN 0-471-71397-X, Hoboken, New Jersey, USA.

Uhlin-Hansen, L., Wik, T., Kjellen, L., Berg, E., Forsdahl, F., Kolset, S.O. (1993) Proteoglycan metabolism in normal and inflammatory human macrophages. *Blood,* Vol. 82, No. 9, (November 1993), pp. 2880–2889, ISSN 0006-4971.

Wijesekara, I., Pangestuti, R., Kim, S.H. (2011). Biological activity and potential health benefits of sulphated polysaccharides from marine algae *Carbohydrate Polymers,* Vol.84, No.1, (February 2011), pp. 14-21, ISSN 0144-8617

Yu, P., Li, N., Liu, X., Zhou, G., Zhang, Q., Li, P. Antihyperlipidemic effects of different molecular weight sulphated polysaccharides from Ulva pertusa (Chlorophyta) (2003). *Pharmacological Research,* Vol.48, No. 6 (2003), pp. 543–549, ISSN: 1043-6618.

Zhang, H.J., Mao, W.J., Fang, F., Li, H.Y., Sun, H.H., Gehen, Y., Qi, X.H., Chemical characteristics and anticoagulant activities of a sulphated polysaccharide and its fragments from Monostroma latissimum, *Carbohydrate Polymers,* Vol.71, No.3, (February 2008), pp. 428-434, ISSN 0144-8617

4

Biomaterials and Epithesis, Our Experience in Maxillo Facial Surgery

G. Fini, L.M. Moricca, A. Leonardi,
S. Buonaccorsi and V. Pellacchia
La Sapienza/ Roma
Italy

1. Introduction

Maxillofacial prosthetics is considered in literature as "... the art and science of anatomic, functional and cosmetic reconstruction, by the use of non-living substitutes, of those regions in the maxillae, mandible and face that are missing or defective..." 1. In the maxillofacial surgery where malformative, oncologic traumatologic pathology and the plastic surgery are treated, the maxillofacial prostheses, in selected cases, can reach a satisfactory therapeutic result from functional, aesthetic, psychologic, and social point of views. In a delicate district, such as the face, where a heavy deficit can determine huge psychologic and social problems, the conventional reconstructive surgery intervenes with reconstructive techniques and with the biomaterials insertion, often insufficient to guarantee the restoration of the harmony of the face. When these conditions are verified, the solution resides in the osteointegration concept and in the application of the epithesis. There are certainly some limits of application of these prostheses, first, the ethics limits: the epithesis constitute in fact an alternative only when the conventional reconstructive surgery cannot be applied, but inside these limits, it is really possible to find an excellent therapeutic resource in patients who cannot undergo surgical interventions. In literature, it is possible to find different kinds of reconstruction of missing body parts by the application of prothesis2.The osteointegration concept was introduced at first time by Professor Branemark in 1960 to describe the "direct structural and functional connection between living bone and the surface of a plant exposed to load, understood as a not static but dynamic process3. According to his school of thought, the technique of positioning of the implant is fundamental, to take place in the most complete precision and to allow the initial stability of one's self. Other elements conditioning the success of the osteointegration are the material of the implant, the form, the areas of the application, and the patient's clinical conditions. The first titanium osteointegration implant was positioned in 1965 in the jaw without dental elements 4;in 1977, implants were positioned in mastoid areas for the application of an acoustic translator. In 1979,implants for the fixation of epithesis of ears, noses, and eyes were positioned. At present, the indication to the position of epithesis as the first choice of treatment is when the conventional reconstructive interventions turn out to be inapplicable or ineffective. The epithesis is a good resolution for the patient because it is not traumatic and has short-time result, removing every psychologic physique obstacle for the inclusion in a normal social life.

2. Our experience

From May 2002 to December 2010, 415 facial prosthesis (1117 implants) have been positioned in our Ephitesy Center. Defects were congenital (N = 142), consequent to trauma (N = 95) and to demolitive surgery for malignant tumors (N = 95), and infection (N = 83). In 40 patients, implants were placed in previously irradiated areas. A total of 1117 titanium implants were placed to support 187 auricular prostheses (bilateral in 29 cases), 126 orbital prostheses, 89 nasal prostheses, and 13 complex midfacial prostheses.

Clinical Case 1U.G., 57-year-old patient, came to our observation with ethmoidal-sphenoidal-orbital-hemimaxillary resection and reconstruction with pectoral flap complicated in the same year by cerebral abscess of Eikenella. The patient was presenting the absence of the skeleton structures and the soft tissues of the third middle of the right emi-face with involvement of the nose and of the hard palate. The pectoral flap was causing deficit in the movements of extent and left rotation of the head. As a consequence of a cerebral ictus and for the detachment of septic carotid plaque embolus, the patient presented with hemiplegy. Heavy deficits were furthermore present to deglutition and masticatory function. The patient was arriving to our observation in order to restore the symmetry of the face and the integrity of the hard palate and to recover the motility of the cervical stroke. A surgical intervention of positioning of epithesis to rebuild the third middle and superior of the face and of the revision of the pectoral flap was therefore planned. Four fixtures with related abutments were placed to support anchoration for the midfacial prosthesis (Figs 3 and 4). In addition, a dental implant was placed in the right tuber maxillae to support a palatal obturator (Fig 5). Finally, a surgical revision of the pectoral flap was performed. Ten months after surgery, a palatal obturator was placed so that it was possible to remove percutaneous endoscopicgastrectomy (PEG).

Clinical Case 2, R.A., a 40-year-old man affected by the Goldenhar syndrome, underwent different reconstructive surgical treatments to restore the normal symmetry of the face soft tissues. The patient came to our center presenting a facial asymmetry characterized by atrophy of the right hemifacial soft tissues, associated to auricular agenesy and to esterior uditive conduct and "anteroposizione" of the left auricular (Figs 6 and 7). Clinical and radiologic examinations with computer tomography dental scan and Telecranium x-ray in 2 projections with cefalometric study were performed to evaluate the bone and the soft tissues. After 1 month, a surgery has been performed to remove the residual cartilage planted in the site corresponding to porous polyethylene prosthesis, positioned during the previous surgical treatment. In addition, 2 fixtures with abutment have been positioned in the right mastoid bone. Then the left auricular was positioned to reestablish the normal structures of the face. In the same surgical time, 2 porous polyethylene prostheses were implanted in the malar region to restore the sagittal diameter of the middle third of the face; then 2 porous polyethylene prostheses were implanted in the mandibular angle, and 1 prosthesis was implanted on the mandibulae, to restore the transversal and sagittal diameter of the third inferior of the face. After 3 months, an auricular prosthesis associated to polyacrylamide implant was positioned in bilateral preauricular area (Figs 8 and 9). Clinical and radiologic follow-up demonstrated a good integration of implants and the biomaterial

Clinical Case 3 A.S., a 51-year-old man affected with posttraumatic anophthalmia, sequelae of left orbit exenteration and reconstruction of the eye socket with a titanium mesh covered by dermo-adipose flap, came to our observation with anophthalmia O.S. and fibrotic scars. Clinical and radiologic examinations with three-dimensional computed tomography were performed to evaluate the bone and the soft tissues (Figs 10 and 11). After the clinical and radiologic evaluation and the patient's agreement, 4 fixtures with corresponding abutments were placed to support the anchor of the orbital epithesis. Nasal and orbital scars were corrected by little flaps (Figs 12 and 13).

Clinical Case 4 F.M., a 61-year-old man, was referred with a nose extirpation for a squamocellular cancer on the nasal tip, involving all nasal structure, 7 years before (Fig 14).The patient and his family declined any kind of reconstructive operative interventions, so the patient underwent nasal movable prosthesis resting. Based on this situation,wehad proposed tohimnasal removable prosthesis fixed with bone paranasal implants. For this reason, the patient had undergone computed tomography scan of the head and neck to study bone density and then 2 implants (4 mm) were placed. Follow-up at 3, 6, and 12 months with clinical visits and computed tomography scan revealed correct implant bone integration (Fig 15).

Clinical Case 5 P.D., a 25-year-old woman, underwent surgical exenteration orbitae because of retinoblastoma. The orbital cavity was restored by temporal muscle flap and dermal-free flap. The patient underwent many reconstructive surgical treatments through the use of fillers of biomaterials in frontal-temporal-cheek side, to reconstitute the anatomic structure. She arrived in our observation with a moving orbital prosthesis (Fig 16). Clinical and radiologic examinations with three-dimensional computed tomography were performed to evaluate the bone and the soft tissues. In accordance with the patient's desire, 3 titanium fixtures with abutments were implanted to position the orbital prosthesis (Fig 17).

Clinical Case 6 M.N., a 56-year-oldwoman,was referred with a partial auricular extirpation for a basocellular cancer on the auricular left elice. The 2/3 superiors of the auricular pavilion have been removed, with a partial deficit of the pavilion itself, which has caused psychologic problems to the patient. In agreement with the patient, a second surgical treatment was performed, modeling porous polyethylene peace with Nagata technique and covered by temporoparietal fascia and dermo-epidermic flap to fill the auricular fault. The biomaterial is not osteointegrated, so it has been removed. For such reason, in agreement with the patient justified strongly to an immediate and no invasive aesthetic rehabilitation, 2 fixtures with abutments have been positioned that support auricular epithesis (Figs 18Y20). The clinical and radiologic follow-up has shown a correct osteointegration of the implants reaching psychologic stability of the patient.

Clinical Case 7 G.B., a 68-year-old woman, with epatotrasplanting and hepatitis C virus has arrived in our observation with a necrotic lesion of the nasal tip resulting to immunosuppressive therapy. She was referring to have noticed the appearance of the necrosy and his progressive growth soon after the end of the therapy. The patient was presenting exposure of the cartilaginous septum with erosion and cutaneous necrosy to the nasal base (Fig 21). Because of the clinical conditions of the patient, a fixture's implant has been made for the positioning of an epithesis in order to obtain an effective reconstruction. Three fixtures with abutments have been applied. A fixture was removed approximately 2 months after the installing because it is not integrated. The other 2 implants seemed to be well supplemented to allow the positioning of the bar that supports the epithesis, but after 2months, 1 fixture has been removed because of missed osteointegration. Therefore, it was decided to position some magnets to anchorage the epithesis (Fig 22).

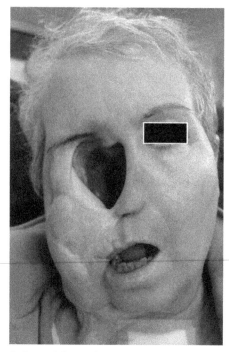

Fig. 1. Preoperative frontal view of the patient.

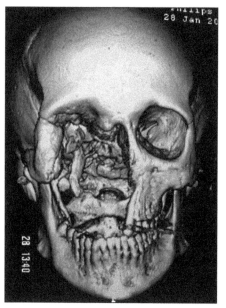

Fig. 2. Preoperative three-dimensional computed. TomographyVfrontal view of the patient.

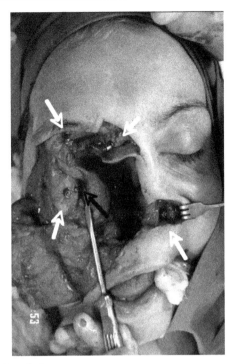

Fig. 3. Intraoperative point of view.

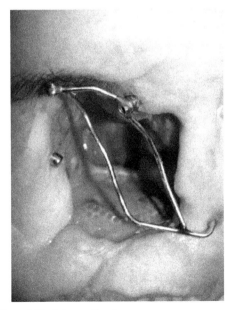

Fig. 4. Anchoration for the midfacial prosthesis.

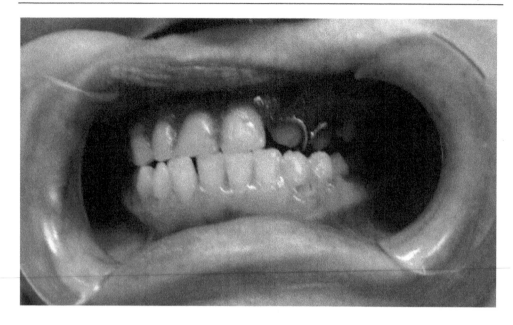

Fig. 5. The palatal obturator.

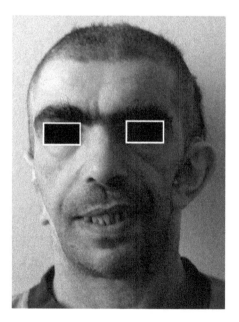

Fig. 6. Preoperative frontal view of the patient.

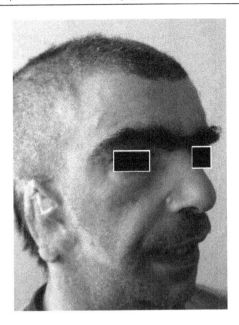

Fig. 7. Preoperative lateral view of the patient.

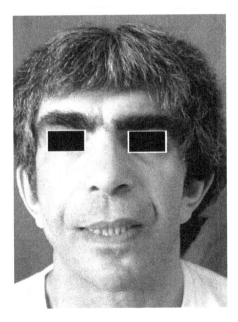

Fig. 8. Postoperative frontal view of the patient.

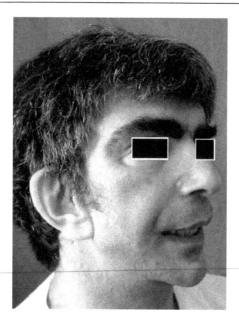

Fig. 9. Postoperative lateral view of the patient.

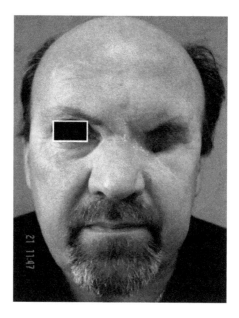

Fig. 10. Preoperative frontal view of the patient.

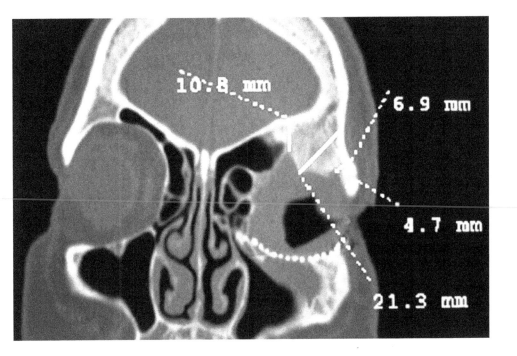

Fig. 11. Preoperative computer tomographyVfrontal view.

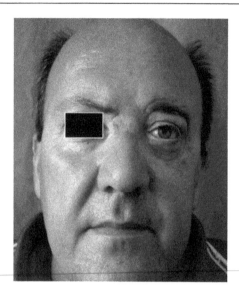

Fig. 12. Postoperative frontal view of the patient.

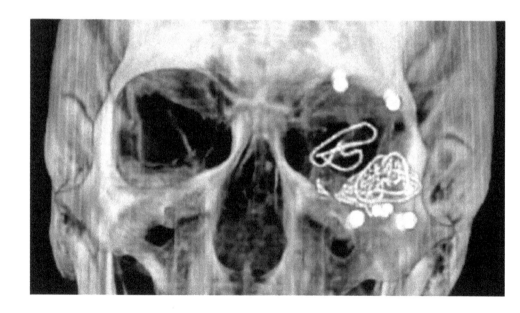

Fig. 13. Postoperative computer tomographyVfrontal view.

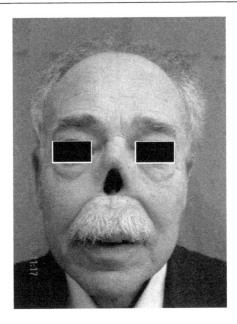

Fig. 14. Preoperative frontal view of the patient.

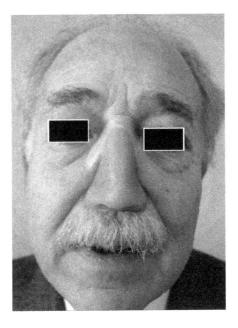

Fig. 15. Postoperative frontal view of the patient.

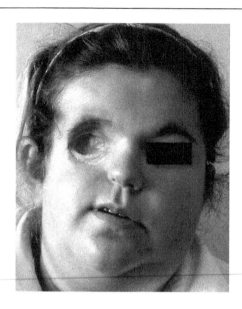

Fig. 16. Preoperative frontal view of the patient.

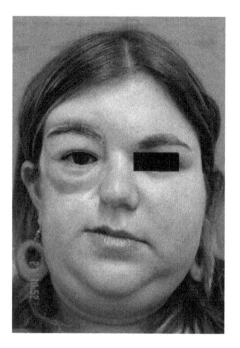

Fig. 17. Postoperative frontal view of the patient.

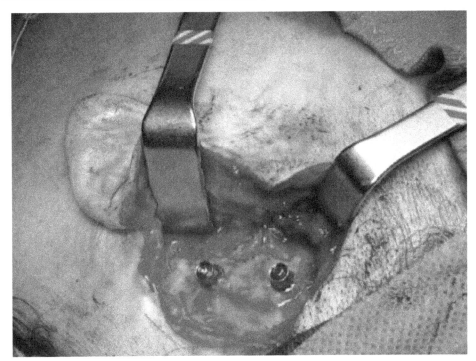

Fig. 18. Fixtures positioning.

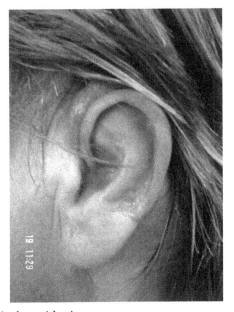

Fig. 19. Patient with auricular epithesis.

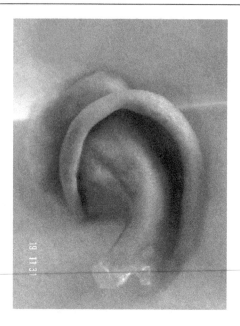

Fig. 20. Auricular epithesis.

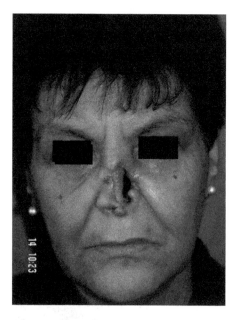

Fig. 21. Preoperative frontal view of the patient.

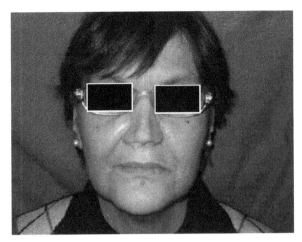

Fig. 22. Postoperative frontal view of the patient.

3. Conclusion

The facial prosthetic rehabilitation is a valid alternative when the conventional reconstructive surgical techniques cannot be applied either because of the psychophysical conditions of the patient or because of an excessive substance loss. The surgical technique with prosthesis has several applications: malformative, infective, traumatic pathology, results of oncologic surgery and radiant therapy, and particular clinical conditions such as diabetes, leukemia, and others. The position of epithesis, as described in the literature5,6 and confirmed by the experience of our epithesis Center, is suitable in selected cases:

• reconstruction with patient's own tissue, which is uneventful or impossible;
• "Reversible" intervention to operate clinically;
• Surveillance in oncologic patients;
• Advanced age or poor health; and poor tissues quality patient's choice

The described technique presents absolute limits such as osteolitic process, leukemia-lymphoma, and terminal cirrhosis and relative limits such as ending life, hygienic deficiency, and psychological refuse. Another important limit is the radiotherapy treatment; the skeletal structure of persons who undergone radiotherapy react to the osteointegration process with a lower success percent. It goes, in fact, to consider that if the combined application of the chemotherapy and radiotherapy treatments with demolitive surgery increases the life on average, the survival of the subject with surgical cancerYablation increases, compromising the quality of life.7 The results of the osteointegration in patients who have underwent chemotherapy are very variable, approximately 60% and 100%.8 In accordance with the literature, we can affirm that the radiotherapy compromises the human tissues, hindering the osteointegration process, when the irradiation is around 5000 Gy. Besides the site and the radiation dose, the time existing between the radiant treatment and the positioning of the implant is another determinant factor for the success of osteointegration process. In particular, 6 months should exist between the term of the radiant treatment and the positioning of the implant period in which the tissue alteration produced by the radiations are in regression. According to the oncologic guideline, it would be more opportune to wait 1 year to avoid the recidivism risk.8 Furthermore, the treatment in the hyperbaric room is effective in the bone life, with higher success percents.8,9 Another fundamental aspect is the epithesis stability,which depends frommany circumstances such as hygienic condition, material quality, and the correctmethod of the epithesis production;when these conditions are respected, the epithesis can resist for 2 years. The application of an epithesis happens with no invasive and immediate results, both fromthe aesthetic and psychologic point of views, allowing to get around with the heavy social insertion problems derived from his facial deformation. The therapeutic iter in the reconstructive treatment with epithesis foresees a dynamic study with few fundamental stages:

- clinical, radiologic, and psychologic evaluation;
- surgical planning;
- positioning of the fixtures;
- templating;
- preparation of the epithesis;
- fixtures; and epithesis exposure.

Beyond the application of bone implants, several retentionmethods are possible: anatomic, exploiting the premade cavity getting to the deficit (ocular epithesis), andmechanical, exploiting outside anchorage strengths (sight glasses) and adhesive, by glue.10 Thanks to the use of the bone implants, it has been able to get around the problems caused by the use of adhesives like decoloration, the precocious deterioration of the epithesis, and inflammatory phenomena of the skin in contact with epithesis' materials. Under the point of view of the aesthetic result, the margins of an epithesis can be easily hidden, and the prosthesis ismore stable, is easy to wear, and keeps under a hygienic point of view. Furthermore, the psychologic appearance should not be neglected because, unlike traditional prosthesis, the epithesis fixed with implants are not considered as an extraneous object, with the consequent improvement of a good quality of life. At present, our experience teaches us that the indication to the position of epithesis as the first choice of treatment is when the conventional reconstructive interventions turn out to be inapplicable or ineffective

4. References

Bulbulian AH. Maxillofacial prosthetics: evolution and practical application in patient rehabilitation. J Prosthet Dent 1965; 15:554Y569

Fini Hatzikiriakos G. Uno sguardo al passato, curiosita` sulle protesi nasali. Il Valsala 1985;61:61Y64

Tjellstrom A. Osteointegrated implants for replacement of absent or defective ears. Clin Plast Surg 1990;17:355Y366

Tjellstrom A, Granstrom G. One stage procedure to establish osteointegration: a zero to five years follow-up report. J Laryngol Otol 1995;109:593Y598

Schaaf NG. Maxillofacial prosthetics and the head and neck cancer patients. Cancer 1984;54:2682Y2690

Labbe´ D, Be´nateau H, Compe`re JF, et al. Implants extra-oraux: indications et contre-indications. Rev Stomatol Chir Maxillofac 2001;102:239Y242

Be´nateau H, Crasson F, Labbe´ D, et al. Implants extra-oraux et irradiation: tendances actuelles. Rev Stomatol Chir Maxillofac 2001;102:266Y269

Granstrom G, Jacobsson M, Tjellstrom A. Titanium implants in irradiated tissue: benefits from hyperbaric oxygenation. Int J Oral Maxillofac Implants 1992;7:15Y25

Markt JC, Lemon JC. Extraoral maxillofacial prosthetic rehabilitation at the M.D. Anderson Cancer Center: a survey of patient attitudes and opinions. J Prosthet Dent 2001;85:608Y613

Tjellstrom A, Granstrom G. One-stage procedure to establish osteointegration: a zero to five years follow-up report. J Laryngol Otol 1995;109:593Y598

Ramires PA, Miccoli MA, Panzarini E, Dini L, Protopapa C. In vitro and in vivo biocompatility of a Polylkymide Hydrogel for soft tissue augmentation. J.Biomed.Mater.Res.Part.2004; 72: 230-238. 2004.

Christensen LC, Breiting VB, Aasted A, Jorgensen A, Kebuladze I. Long-Term effects ofpolyacrylamide hydrogel in human breast tissue. Plast.Reconstr.Surg.2003; 11:1883-1890

Rees TD, Ashlet FL, Delgado JP. Silicone fluid injectons for facial atrophy: a 10 years study.Plast.Reconstr.Surgery. 1985;52: 118-125.

Greenwald AS,bBoden SD, goldberg VM et al.Bone graft substitutes: Facts, fictions and applications. J. Bone Joint Surg Am 2001; 83-A:S98-S103

Bauer TW, Muschler GF. Bone Graft Materials. An overview of The basic scienze. Clin Orthop. 2000; 371:10-27.

Oka Y, Ikeda M. Treatment of severe osteochondritis dissecans of the elbow using osteochondral grafts from a rib. J Bone Joint Surg Am. 2001 83-b: 738-739

Stevenson S. Biology of bone grafts. Orthoped Clin North Am. 1999; 30:543-552.

Skowronski PP, An Yh. Bone graft materials in orthopaedics.MUSC Orthopaed J. 2003;6:58-66

Betz RR. Limitation of autograft and allograft: new synthetic solutions. Orthopedics 2002;25(Suppl):S561-S560

Hollinger JO, Mark DE, Goco P, et al. A comparison of four particulate bone derivatives. Clin

Pietrzak WS, MillervSD, Kucharzyk DW, et al. Demineralized bone graftvformulations: Design, development, and a novel exemple. Proceedings of the Pittsburg BonevSymposium, Pittsburgh, PA, August 19-23,"003,557-575.

Davy DT.Biomechanical issues in bone transplantation. Orthop Clin North Am.1999;30:553-56 science and technology for the Craniomaxillofacial Surgeon.J Craniofac. Surg. 005(16);6:981-988.15

Mauriello JA, McShane R, Voglino J. Use Vicryl(Polyglactina 910) mesh implant for correcting enophtalmos A study of 16 patients. Ophthal Plast Reconstr Surg 1990; 6:247-251.

S. Ozturk, M Sengezer, S Isik et all. Long Term Outcomes of Ultra Thin Porous Polyethylene Implants used for Reconstruction of Orbital Floor Defects. J Craniofac Surg. 2005 (16) 6:973-977.

S. Buelow, D Heimburg, N. Pallua. Efficacy ad Safety of Polycrylamide Hydrogel for FacialSoft-Tissue Augmentation. Plast Reconstr Surg 2005 (15);1137-1146.

KW Broder, SR Cohen. An Overview of Permanent and Semipermanent Fillers. Plast Reconstr Surg 2006 118(3 suppl.) S7_S14

Biomaterial Implantation in Facial Esthetic Diseases: Ultrasonography Monitor Follow-Up
Elena Indrizzi, MDS, Luca Maria Moricca, MD, Valentina Pellacchia, MD, Alessandra Leonardi, MD, Sara Buonaccorsi, MD, Giuseppina Fini, MDS, PhD. The Journal of Craniofacial Surgery –Vol.19, N. 4 -July 2008

Human Dentin as Novel Biomaterial for Bone Regeneration

Masaru Murata[1], Toshiyuki Akazawa[2], Masaharu Mitsugi[3],
In-Woong Um[4], Kyung-Wook Kim[5] and Young-Kyun Kim[6]
[1]Health Sciences University of Hokkaido,
[2]Hokkaido Organization,
[3]Takamatsu Oral and Maxillofacial Surgery
[4]Tooth Bank Co. Ltd,
[5]Dankook University,
[6]Seoul National University Bundang Hospital,
[1,2,3]Japan
[4,5,6]Korea

1. Introduction

Human dentin autograft was reported in 2003 as a first clinical case (Murata et al., 2003), while human bone autograft was done in 1820. There was a long-long time lag between the autografts of dentin and bone. In 2009, Korea Tooth Bank was established in Seoul for the processing of the tooth-derived materials in Seoul, and an innovative medical service has begun for bone regeneration. Recently, the tooth-derived materials have been becoming a realistic alternative to bone grafting.

The regeneration of lost-parts of the skeleton has been generally carried out with fresh, autogenous bone as a gold standard. To obviate the need for harvesting of grafts and thus, to avoid morbidity resulting from it, the researches for bone substitutes (Kuboki et al., 1995; Asahina et al., 1997; Takaoka et al., 1991; Artzi et al., 2004; Kim et al., 2010) or bone production via bio-engineering have begun (Wozney et al., 1988; Wang et al., 1990; Murata et al., 1999). In the regenerative field, there is a medical need for biomaterials that both allow for bone formation and also gradually absorb as to be replaced by bone. Non-absorbable materials are never replaced by bone and thus, reveal chronic inflammation in tissues as foreign bodies.

As bone and dentin consist of fluid (10%), collagen (20%) and hydroxyapatite (70%) in weight volume, our attention for biomaterials is collagenous and ceramic materials (Murata et al., 2000; Murata et al., 2002; Akazawa et al., 2006; Murata et al., 2007). Generally, extracted teeth have been discarded as infective medical dusts in the world. We have thought the non-functional teeth as native resource for self and family (Fig. 1). Therefore, we noticed on bone-inductive, absorbable properties of dentin, and have been studying a medical recycle of human teeth as a novel graft material for bone regeneration in Japan and Korea (Akazawa et al. 2007; Kim et al. 2010). Biomaterial science should support and develop the advanced regenerative therapy using enamel and dentin matrix for patients in the near future.

In this chapter, human dentin will be introduced as novel biomaterial and also as carrier matrix of the recombinant human bone morphogenetic protein-2 (BMP-2) delivery for bone engineering.

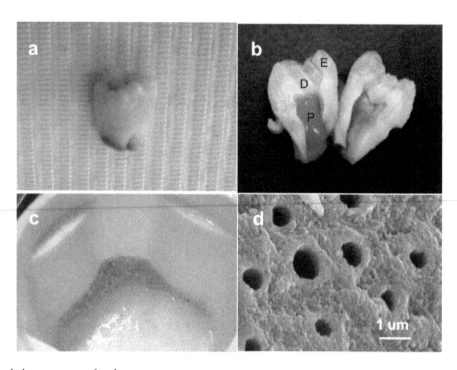

a: whole appearance of molar.
b: divided tooth (E; enamel, D; dentin, P; pulp).
c: crushed tooth granules.
d: SEM photograph of calcified dentin after crushing and washing. Note; dentinal tubes

Fig. 1. Human wisdom tooth

2. Bone induction of human dentin

In 1967, bone-inducing property in rabbit dentin was confirmed in the intramuscular pockets (Yeoman &Urist, 1967; Bang & Urist, 1967), after the discovery of bone induction by rabbit demineralized bone matrix (DBM) in 1965 (Urist, 1965). The rabbit studies reported that completely demineralized dentin matrix (DDM) induced bone at 4 weeks, while non-demineralized dentin (so-called, calcified dentin) induced bone at 8-12 weeks after implantation (Yeoman & Urist, 1967). In our study, human DDM including small patches of cementum derived from wisdom teeth, and human DBM derived from adult femur induced bone and cartilage independently in the subcutaneous tissues at 4 weeks (Murata et al., 2010a). The delayed inductive properties of the calcified dentin and bone may be related to the inhibition of BMP-release by the apatite crystals. Highly calcified tissues such as cortical bone and dentin are not earlier in osteoinduction and bone fornmation than spongy bone, decalcified bone (DBM), and decalcified dentin (DDM) (Huggins et al., 1970).

Dentin and bone are mineralized tissues and almost similar in chemical components. Both DDM and DBM are composed of predominantly type I collagen (95%) and the remaining as non-collagenous proteins including small amount of growth factors (Finkelman et al., 1990). In other words, DDM and DBM can be defined as acid-insoluble collagen binding bone morphogenetic proteins (BMPs), which are member of transforming growth factor-beta (TGF-β) super-family. BMPs were discovered from bone matrix (Urist, 1965; Sampath & Reddi., 1983), and had bone-inducing property in non-skeletal site (Murata et al., 1998). Animal dentin-derived BMPs were extracted with 4M guanidine HCl, and partially purified from rat, rabbit, and bovine (Butler et al., 1977; Urist & Mizutani, 1982; Kawai & Urist, 1989; Bessho et al, 1990). In addition, the concentration of TGF-β, Insulin growth factor-I (IGF-I) and Insulin growth factor-II (IGF-II) were detected in human dentin (DDM). Briefly, the three growth factors were measured in the following concentration (ng/μg 4M guanidine hydrochloride-EDTA protein): TGF-β (0.017), IGF-I (0.06) and IGF-II (0.52). All 3 growth factors were present in concentrations lower than that in human bone (Finkelman et al., 1990). Recently, both mature and immature types of BMP-2 were detected in human dentin and dental pulps (Ito et al., 2008).

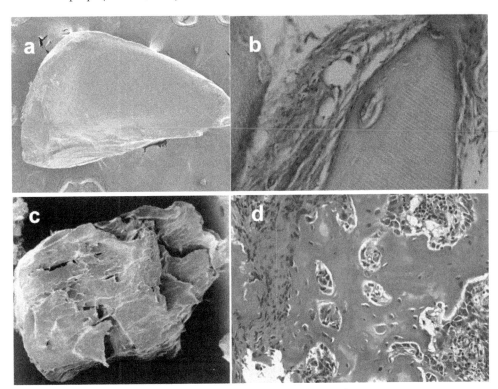

a: SEM of DDM (granule size: 0.5mm), Note: smooth surface and no crack.
b: bone induction by DDM at 4 weeks.
c: SEM of DBM (granule size: 0.5mm), Note: micro-cracks and spaces of blood vessels.
d: bone induction by DBM at 4 weeks.

Fig. 2. Dematerialized dentin matrix (DDM) and dematerialized bone matrix (DBM)

Even after the demineralization of dentin, active types of BMPs bind collagen-rich matrices, similar to bone (Urist et al., 1973). The decalcified dentin (DDM) was known to be more active bone-inducing matrix than the calcified dentin (Yeoman & Urist, 1967), and roll type of decalcified dentin membrane revealed better activity of bone induction (Inoue et al., 1986).

Very interestingly, the demineralized treatment for bone and dentin increased their osteoinductivity and decreased their antigenesity (Reddi, 1974). These facts are scientifically very important for the processing procedures of hard tissue-derived graft materials (Kim et al., 2010; Murata et al, 2010a).

The acid-insoluble dentin matrix (DDM) after demineralization is an organic, absorbable material with original dentin structures. Human DDM, prepared from vital teeth-origin, were implanted into the subcutaneous tissue in 4 week-old nude mice, deficient in immunogenic reactions. The DDM induced bone and cartilage independently at 4 weeks after the subcutaneous implantation, similar to human DBM (Murata et al., 2010b). The independent differentiation of bone and cartilage was compatible to our previous study using ceramic and collagen combined with BMPs (Murata et al., 1998). The acid-insoluble collagen, DBM and DDM, possess the ability to coagulate platelet-free heparinized, citrated, and oxalated blood plasmas (Huggins & Reddi., 1973). Clotting constituents become denatured in contact with the insoluble coagulant proteins. The coagulation action of blood plasma by DBM and DDM should become advantageous for surgical operations. Collagenous materials has been commercially available as medical uses for more 30 years.

3. Clinical study of human dentin

3.1 Case 1: Bone augmentation, 48 year-old man

First clinical study was reported at 81st IADR conference, Sweden in 2003 that DDM autograft had succeeded for bone augmentation (Murata et al., 2003).

The aim of this pioneering study is to observe new bone formation in the tissues obtained from the dental implant-placed region after the DDM graft for sinus lifting.

Patient

A 48-year-old male presented with missing teeth (#24-#26, #45-#47). Clinical examinations revealed an atrophied upper jaw in the region (Fig. 3,4). His medical history was unremarkable.

Surgical procedure 1

Four teeth (#17,#18,#25,#28) were extracted and 2 molars (#17,#18) were stocked at -80℃ for DDM. His right occlusion was restored using dental implants as the first clinical step (Fig. 4b).

Preparations of DDM

The autogenous DDM were obtained from non-functional vital teeth (#17, #18) (Fig. 4a). The molars were crushed by hand-made under the cooling with liquid nitrogen. The crushed tooth granules were decalcified completely in 0.6N HCl solution. The DDM granules including cementum were extensively rinsed in cold distilled water, and then freeze-dried (Murata et al., 2010a).

Surgical procedure 2

Sinus lifting procedure was done using autogenous dry DDM for bone augmentation (Fig. 3). At 5 months after the operation, 3 fixtures (FLIALIT-2®, FRIADENT) were implanted into the augmented bone under local anesthesia (Fig. 4c). At the same time, bone biopsy was carried out for the tissue observation (Fig. 4d).

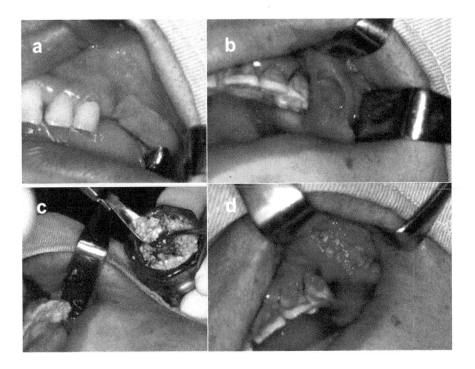

a: intraoral initial view (before operation), Note: 3 missing teeth and atrophied maxilla.
b: oval shaped window
c: autogenous DDM derived from 2 molars
d: view just after DDM autograft

Fig. 3. Case 1: DDM autograft for sinus lifting, 48 year-old man

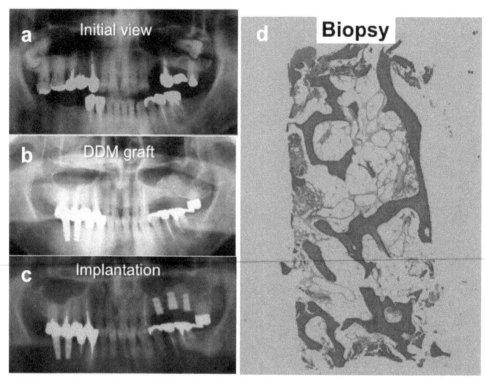

a: initial view, b: 4 months after DDM graft, c: dental implant placement, d: mature bone with marrow

Fig. 4. X-ray photography and bone biopsy

Results and discussion

The biopsy tissue showed that mature bone was interconnected with the remained DDM granules (Fig. 4d). We found that DDM facilitated its adaption of the grafted site and was slowly absorbed as new bone began to form.

Conclusion

This patient was successfully restored with the dental implants after the DDM autograft. These results demonstrated that autogenous dentin could be recycled as an innovative biomaterial.

3.2 Case 2: Bone regeneration, 58 year-old woman

Patient

A 58-year-old female presented with missing teeth (#12-#22). A clinical examination revealed an atrophied upper jaw in the section. Her medical history was unremarkable.

Preparations of DDM

The autogenous DDM were obtained from a non-functional vital tooth (#17). The second molar was crushed with saline ice by our newly developed tooth- mill (DENTMILL®, Tokyo Iken Co., Ltd) at 12000rpm for 30 sec (Fig. 5). Briefly, vessel and blade were made in ZrO_2 , which have gained the approval of Food and Drug Administration (FDA) for human use. The ZrO₂ ceramics were fabricated by sintering at 1400°C for 2 h after the slip casting of the mixture of ZrO_2 powder and distilled water (Fig. 5a). As the results of characteristics analyses of ZrO_2 objects, the contraction rate, the relative density, and the bending strength were 21%, 99%, and 400MPa, respectively. The automatic mill could crush a tooth and/or a cortical bone block (1x1x1cm³) under the condition of cooling using saline ice blocks (1cm³) (Fig. 5b). The crushed tooth granules were decalcified completely in 0.026N HNO_3 solution for 20 min. The DDM granules including cementum were extensively rinsed in cold distilled water (Fig. 5e), (Murata et al., 2009; Murata et al., 2010a).

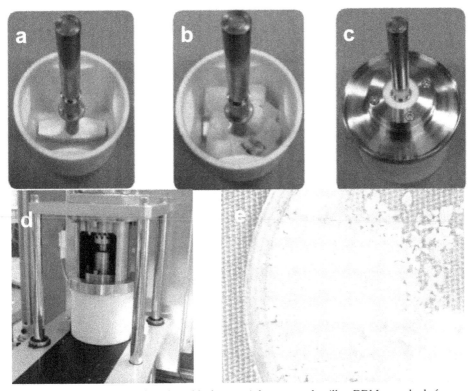

a: ZrO₂ vessel and blade, b: tooth with ice blocks, c: stainless cover, d: mill, e: DDM granules before clinical use.

Fig. 5. Preparation of DDM using automatic tooth mill (DENTMILL®, Tokyo Iken)

Surgical procedure

Splitting osteotomy and cortical perforations were performed in the atrophied jaw and the autogenous DDM were transplanted to the treated bone in 2006 (Fig. 6a,b,c). At 4 months after the operation, 3 same fixtures (Synchro-steppted screw type: diameter; 3.4mm, length; 11mm, FLIALIT-2® , FRIADENT) were implanted into the augmented bone under local anesthesia   (Fig. 6b). At the same time, bone biopsy was carried out for the tissue observation.

Results and discussion

The biopsy tissue showed that DDM granules were received to host and the biological width (4-6mm) was acquired. The DDM residues were partially observed during the implant placement. Bone biopsy revealed the DDM were remodeled by bone at 4 months. This patient was successfully restored with the dental implants after the DDM autograft (Fig. 6d). Though animal-derived atelocollagens have been generally used as medical materials, autogenous decalcified dentin is a highly insoluble collagenous matrix and a safe biomaterial.

Conclusion

Human DDM granules from vital teeth are collagenous matrics with osteoinductive potency, and the human dentin can be recycled as autogenous biomaterials for local bone engineering.

Case 1 and 2 were approved by the Ethical Committee in the Health Sciences University of Hokkaido. All subjects enrolled in this research have responded to an Informed Consent which has been approved by my Institutional Committee on Human Research and that this protocol has been found acceptable by them.

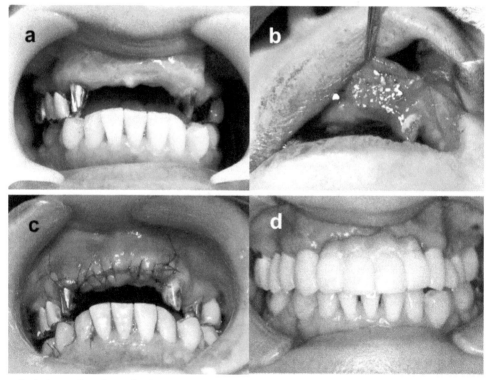

a: 4 missing teeth and atrophied upper maxilla b: DDM autograft before suture c: just after operation d: final view after prosthetic restoration using dental implantation

Fig. 6. Case 2: Bone regeneration, 58 year-old woman

4. Dentin scaffold for recombinant human BMP-2

4.1 Recombinant human BMP products

BMP-2, 4, and 7 are strong accelerating factors of bone induction. Currently, BMP-2 and BMP-7 have been shown in clinical studies to be beneficial in the therapy of a variety of bone-related conditions including delayed union and non-union. BMP-2 (Medtronic Co.Ltd.) and BMP-7 (Stryker Biotech Co.Ltd.) have received Food and Drug Administration (FDA) approval for human clinical uses (fractures of long bones, inter-vertebral disk regeneration), by delivery in purified collagen matrix or ceramics. Moreover, the BMP-2 product has been approved for certain dental applications. BMP-7 has also found use in the treatment of chronic kidney disease. In 2002, Curis licensed BMP-7 to Ortho Biotech Products, a subsidiary of Johnson & Johnson.

4.2 Acceleration of bone induction by BMP2 in human DDM scaffold

The aim of the following study was to estimate the increase of the bone-inductive potency by DDM combined with BMP-2 in rat subcutaneous tissues.

Composition of BMP-2 solution and DDM

One hundred micro-liter of recombinant human BMP-2 solution (0.0, 0.5, 1.0, 2.0, 5.0µg of BMP-2) was mixed with 70 mg of human DDM in a sterilized syringe. The composite was called as the BMP-2/DDM. The DDM alone with 100µl of PBS was also prepared as a BMP-free control.

Bioassay in rats

Wistar rats (male, 4 week-old) were subjected to intraperitoneal anesthesia and incisions were added to the back skin under the sterile conditions. Each animal received three BMP-containing composites (BMP-2/DDM) and one BMP-free control (DDM alone). The implanted materials were removed at 3 weeks after implantation, and prepared for histomorphological examinations. All procedures were followed the Guidelines in Health Sciences University of Hokkaido for Experiments on Animals.

Histological findings and Morphometric analysis at 3 weeks

In the BMP-2 (5.0µg)/DDM (70mg) group, bone with hematopoietic bone marrow developed extensively at 3 weeks. Chondrocytes were found only in the BMP-2 (0.5, 1.0µg)/DDM groups (Table 1). The BMP-2 (2.0, 5.0µg)/DDM groups accelarated bone induction predominantly (Fig. 7). In the DDM alone group, mesenchymal tissue was seen between DDM particles, and hard tissue induction was not observed at 3 weeks (Fig. 8). Morphometric analysis demonstrated that the volume of the induced bone and marrow increased at BMP-2 dose-dependent manner, while the DDM decreased at the dose-dependent (Table 1). Briefly, the volume of the bone and marrow in BMP-2 (1.0µg)/DDM and BMP-2 (5.0µg)/DDM showed 3.7% and 26.3%, respectively. BMP-2 (0.5µg)/DDM showed 0.0% and 4.0% in the volume of bone and cartilage, respectively.

Conclusion

BMP-2 strongly accelerated bone formation in the DDM carrier system. DDM never inhibited BMP-2 activity and revealed better release profile of BMP-2. These results indicate that human recycled DDM are unique, absorbable matrix with osteoinductivity and the DDM should be an effective graft material as a carrier of BMP-2 delivering and a scaffold for bone-forming cells for bone engineering.

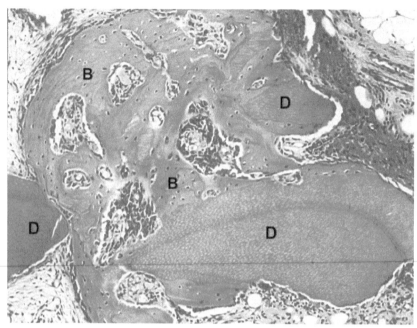

Induced bone (B) bridging between DDM (D) granules. Note: active osteoblast differentiation.

Fig. 7. Photograph in BMP-2 (5.0μg)/DDM (70mg) at 3 weeks

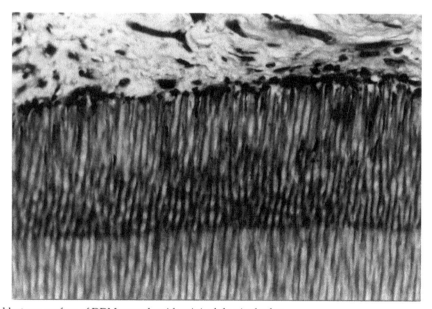

Fibroblasts on surface of DDM granule with original dentinal tubes.

Fig. 8. Photograph in DDM (70mg) alone at 3 weeks

	Dose of BMP-2 (μg)				
	0	0.5	1	2	5
bone	0	0	3.7 ± 1.41	7.4 ± 0.94	20.3 ± 4.64
cartilage	0	4.0 ± 0.81	2.3 ± 0.47	0	0
bone marrow	0	0	0	0	6.0 ± 1.63
DDM	57.0 ± 0.81	43.3 ± 3.39	41.0 ± 2.16	40.3 ± 1.69	37.0 ± 0.81
mesenchymal tissue	40.7 ± 0.94	49.0 ± 5.09	48.0 ± 3.85	46.0 ± 2.16	32.7 ± 5.73
connective tissue	2.3 ± 0.47	3.7 ± 1.24	5.0 ± 0.47	6.3 ± 0.47	4.0 ± 0.81
All tissue: 100 % , values: mean ± SD , N: 9, Explanted time: 3 weeks					

The volume of bone and marrow showing a dose-dependent increase.
The volume of DDM showing a dose-dependent decrease.

Table 1. Morphometry of BMP-2 dose-dependent study.

5. Material science for patients in the near future

Biomaterials have had a major impact on the regenerative medicine and patient care for improving the quality of lives of human.

We have been challenging to be able to develop bioabsorbable materials, harmonized with living body, especially bone remodelling, using an innovative supersonic and acid-etching technology (Akazawa et al. 2010). Implanted biomaterials first contact to body fluid and cells. Human cells never live in dry condition. Generally, organ and tissue have interconnected porous structure for dynamic flow of body fluid. Material walls inhibit the body fluid permeation and the cell invasion. Therefore, we focused on the permeability of body fluid into the bulk of materials and the biomimetic structure for the living and working cells (Murata et al., 2007). Body fluid can permeate into collagenous materials such as DDM and DBM. Novel DDM material contains native growth factors, and adsorbs several proteins derived from body fluid. In addition, DDM with RGD sequences supports mesenchymal cell adhesion as anchorage matrix.

Most importantly, material scientists, engineers, and doctors must work together and cooperate as professionals for the development of functional materials and for the present and future of all patients.

6. References

Akazawa, T., Murata, M., Sasaki, T., Tazaki, J., Kobayashi, M., Kanno, T., Matsushima, K., Itabashi, K., & Arisue, M. (2005). Bio-absorption and osteoinduction innovation of bone morphogenetic protein-supported functionally graded apatites originated from cattle bone. *J Am Ceram Soc*, 88.,12., 3545-3548.

Akazawa, T., Murata, M., Sasaki, T., Tazaki, J., Kobayashi, M., Kanno, T., Matsushima, K., & Arisue, M. (2006). Biodegradation and bioabsorption innovation of the functionally graded cattle-bone-originated apatite with blood compatibility. *J Biomed Mater Res*, 76A., 1., 44-51.

Akazawa, T., Murata, M., Hino, J., Nakamura, K., Tazaki, J., Kikuchi, M., & Arisue, M. (2007). Materials design and application of demineralized dentin/apatite composite granules derived from human teeth. *Archives of Bioceramics Research*, 7., 25-28.

Akazawa, T., Murata, M., Takahata, M., Xianjun, D., Abe, Y., Nakamura, K., Hino, J., Tazaki, J., Ito, K., Ito, M., Iwasaki, N., Minami, A., Nakajima, T., & Sakamoto, M. (2010). Characterization of microstructure and bio-absorption of the hydroxyapatite ceramics modified by a partial dissolution-precipitation technique using supersonic treatment. *Journal of the Ceramic Society of Japan*, 118., 6., 535-540.

Asahina, I., Watanabe, M., Sakurai, N., Mori, M., & Enomoto, S. (1997). Repair of bone defect in primate mandible using a bone morphogenetic protein (BMP)-hydroxyapatite-collagen composite. *J Med Dent Sci.*, 44., 3., 63-70.

Artzi, Z., Weinreb, M., Givol, N., Rohrer, MD., Nemcovsky, CE., Prasad, HS., & Tal, H. (2004). Biomaterial resorption rate and healing site morphology of inorganic bovine bone and beta-tricalcium phosphate in the canine: a 24-month longitudinal histologic study and morphometric analysis. *Int J Oral Maxillofac Implants.*, 19., 3., 357-368.

Bang, G. & Urist, MR. (1967). Bone induction in excavation chambers in matrix of decalcified dentin. *Arch Surg*, 94., 6., 781-789.

Bessho, K., Tagawa, T., & Murata, M. (1990). Purification of rabbit bone morphogenetic protein derived from bone, dentin, and wound tissue after tooth extraction. *J Oral Maxillofac Surg*, 48., 162-169.

Butler, WT., Mikulski, A., Urist, MR., Bridges, G., & Uyeno, S. (1977). Noncollagenous proteins of a rat dentin matrix possessing bone morphogenetic activity. *J Dent Res*, 56., 228-232.

Finkelman, RD., Mohan, S., Jennings, JC., Taylor, AK., Jepsen, S., & Baylink, DJ. (1990). Quantitation of growth factors IGF-I, SGF/IGF-II, and TGF-beta in human dentin. *J Bone Miner Res.*, 5., 7., 717-23.

Huggins, C., Wiseman, S., & Reddi, AH. (1970). Transformation of fibroblasts by allogeneic and xenogeneic transplants of demineralized tooth and bone. *J Exp Med*, 132., 1250-1258.

Huggins, CB., & Reddi, AH. (1973). Coagulation of blood plasma of guinea pig by the bone matrix. *Proc Natl Acad Sci U S A.*, 70., 3., 929-33.

Inoue, T., Deporter, DA., & Melcher, AH. (1986). Induction of chondrogenesis in muscle, skin, bone marrow, and periodontal ligament by demineralized dentin and bone matrix in vivo and in vitro. *J Dent Res*, 65., 12-22.

Ito, K., Arakawa, T., Murata, M., Tazaki, J., Takuma, T., & Arisue, M. (2008). Analysis of bone morphogenetic protein in human dental pulp tissues. *Archives of Bioceramics Research*, 8., 166-169.

Kawai, T., & Urist, MD. (1989). Bovine tooth-derived bone morphogenetic protein. *J Dent Res*, 68., 1069-1074.

Kim, YK., Kim, SG., Byeon, JH., Lee, HJ., Um, IU., Lim, SC., & Kim, SY. (2010). Development of a novel bone grafting material using autogenous teeth. *Oral Surg Oral Med Oral Pathol Oral Radiol Endod.*, 109., 4., 496-503.

Kuboki, Y., Saito, T., Murata, M., Takita, H., Mizuno, M., Inoue, M., Nagai, N. & Poole, R. (1995). Two distinctive BMP-carriers induce zonal chondrogenesis and membranous ossification, respectively; geometrical factors of matrices for cell-differentiation. *Connective Tissue Research*, 31., 1-8.

Murata, M., Inoue, M., Arisue, M., Kuboki, Y., & Nagai, N. (1998). Carrier-dependency of cellular differentiation induced by bone morphogenetic protein (BMP) in ectopic sites. *Int J Oral Maxillofac Surg*, 27., 391-396.

Murata, M., Huang, BZ., Shibata, T., Imai, S., Nagai, N., & Arisue, M. (1999). Bone augmentation by recombinant human BMP-2 and collagen on adult rat parietal bone. *Int J Oral Maxillofac Surg*, 28., 232-237.

Murata, M., Maki, F., Sato, D., Shibata, T., & Arisue, M. (2000). Bone augmentation by onlay implant using recombinant human BMP-2 and collagen on adult rat skull without periosteum. *Clin Oral Impl Res*, 11., 289-295.

Murata, M., Arisue, M., Sato, D., Sasaki, T., Shibata, T., & Kuboki, Y. (2002). Bone induction in subcutaneous tissue in rats by a newly developed DNA-coated atelocollagen and bone morphogenetic protein. *Br J Oral Maxillofac Surg*, 40., 131-135.

Murata, M. (2003). Autogenous demineralized dentin matrix for maxillary sinus augmentation in human. The first clinical report. 81th International Association for Dental Research , Geteburg, Sweden, 2003, June.

Murata, M., Akazawa, T., Tazaki, J., Ito, K., Sasaki, T., Yamamoto, M., Tabata, Y., & Arisue, M. (2007). Blood permeability of a novel ceramic scaffold for bone morphogenetic protein-2. *J Biomed Mater Res*, 81B., 2., 469-475.

Murata, M., Akazawa, T., Tazaki, J., Ito, K., Hino, J., Kamiura, Y., Kumazawa, R., & Arisue, M. (2009). Human Dentin autograft for bone regeneration - Automatic pulverizing machine and biopsy –. *Bioceramics 22*, 22., 745-748.

Murata, M., Kawai, T., Kawakami, T., Akazawa, T., Tazaki, J., Ito, K., Kusano, K., & Arisue, M. (2010a). Human acid-insoluble dentin with BMP-2 accelerates bone induction in subcutaneous and intramuscular tissues. *Journal of the Ceramic Society of Japan*, 118., 6., 438-441.

Murata, M., Akazawa, T., Takahata, M., Ito, M., Tazaki, J., Hino, J., Nakamura, K., Iwasaki, N., Shibata,T., & Arisue, M. (2010b). Bone induction of human tooth and bone crushed by newly developed automatic mill. *Journal of the Ceramic Society of Japan*, 118., 6., 434-437.

Reddi, AH. (1974). Bone matrix in the solid state:geometric influence on differentiation of fibroblasts. *Adv Biol Med Phys*, 15., 1-18.

Sampath, TK., & Reddi, AH. (1983). Homology of bone-inductive proteins from human, monkey, bovine, and rat extracellular matrix. *Proc Natl Acad Sci USA*, 80., 6591-6595.

Takaoka, K., Koezuka, M. & Nakahara, H. (1991). Telopeptide-depleted bovine skin collagen as a carrier for bone morphogenetic protein. *Journal of Orthopaedic Research*, 9., 902-907.

Urist, MR. (1965). Bone: Formation by autoinduction. *Science*, 150., 893-899.

Urist, MR., Iwata, H., Ceccotti, PL., Dorfman, RL., Boyd, SD., McDowell, RM., & Chien, C. (1973). Bone morphogenesis in implants of insoluble bone gelatin. *Proc Nat Acad Sci USA*, 70., 3511-3515.

Urist, MR., Mizutani, H., Conover, MA., Lietze, A., & Finerman, GA. (1982) Dentin, bone, and osteosarcoma tissue bone morphogenetic proteins. *Prog Clin Biol Res*, 101., 61-81.

Wang, EA., Rosen, V., D'alesandro, JS., Bauduy, M., Coredes, P., Harada, T., Israel, DI., Hewick, RM., Kerns, KM., La Pan, P., Luxenberg, DP., Mc Quaid, D., Moutsatsos,

IK., Nove, J., & Wozney, JM. (1990). Recombinant human bone morphogenetic protein induces bone formation. *Proc Natl Acad Sci USA* 87., 2220-2224.

Wozney, JM., Rosen, V., Celeste, AJ., Mitsock, LM., Whitters, MJ., Kriz, RW., Hewick, RM., & Wang, EA. (1988). Novel regulators of bone formation: molecular clones and activities. *Science,* 242., 1528-1534.

Yeomans, JD. & Urist, MR. (1967). Bone induction by decalcified dentine implanted into oral, osseous and muscle tissues. *Arch Oral Biol,* 12., 999-1008.

6

Silanization with APTES for Controlling the Interactions Between Stainless Steel and Biocomponents: Reality vs Expectation

Jessem Landoulsi[1], Michel J. Genet[2], Karim El Kirat[3],
Caroline Richard[4], Sylviane Pulvin[5] and Paul G. Rouxhet[2]
[1]*Laboratoire de Réactivité de Surface,*
Université Pierre & Marie Curie -Paris VI,
[2]*Institute of Condensed Matter and Nanosciences – Bio & Soft Matter,*
Université Catholique de Louvain,
[3]*Laboratoire de Biomécanique et Bioingénierie,*
[4]*Laboratoire Roberval,*
[5]*Génie Enzymatique et Cellulaire,*
Université de Technologie de Compiègne,
[1,3,4,5]*France*
[2]*Belgium*

1. Introduction

The surface of biomaterials is frequently chemically modified with the aim to modify the physicochemical properties (hydrophobicity, electrical charge, solvation) which control the interactions with biomolecules and consequently with cell surfaces, or to retain biochemical entities which are specifically recognized by the cells (Williams, 2010). Regarding inorganic materials, widespread procedures involve self-assembly of alcane thiols on gold, silver, copper or platinum (Wink et al., 1997). However, these substrates have limited interest in biomedical applications. Other procedures consist in grafting organosilanes on silica and other metal oxides (Weetall, 1993). The use of silane coupling agents has been reported in various biomaterials researches, such as surface modification of titanium (Nanci et al., 1998), natural fiber/polymer composites (Xie et al., 2010) or dental ceramics (Matinlinna et al., 2004; Matinlinna & Vallittu, 2007).

The silanization reaction at interfaces is complex and there is still considerable debate on the retention mechanisms and on the organization of the interface (Gooding & Ciampi, 2011; Haensch et al., 2010, Suzuki & Ishida, 1996). Depending on the nature of reactive moieties bound to Si in the silane (typically Cl or alkoxy group) and their number, and on the reaction conditions (particularly the presence of water), the relative importance of covalent binding to the surface, oligomerization, polymerization along the surface plane, three-dimensional polymerization may possibly vary. The efficiency of the surface modification is often demonstrated by its influence on biochemical or biological activity. However the nature of the interface produced is difficult to characterize, which limits the guidelines available to improve the procedures. Moreover organic contaminants are always present on high energy solids. They are mainly of hydrocarbon nature and are readily adsorbed from surrounding air or in surface analysis spectrometers (Caillou et al., 2008; Landoulsi et al., 2008a). The possible influence of contaminants on the silanization process and product is usually not considered. In the case of silicon wafer silanized with 3-[methoxy(polyethyleneoxy)]propyl trimethoxysilane and trichlorosilane in organic solvents under a controlled atmosphere, the surface obtained was described as a 1 to 2 nm thick grafted silane layer covered by a thin layer of adventitious contaminants, suggesting that contamination was posterior to the silanization reaction. On the other hand, the silane layer was not stable in phosphate buffered saline at 37°C (Dekeyser et al., 2008).

Aminopropylalkoxysilanes are attractive for surface modification (Plueddemann, 1991), as their bifunctional nature is expected to offer the possibility of covalently attaching a biomolecule, either directly or through a linker. 3-Aminopropyl(triethoxysilane) (APTES) is one of the most frequently used organosilane agents for the preparation of amine-terminated films (Asenath Smith & Chen, 2008; Howarter & Youngblood, 2006; Kim et al., 2009a; Lapin & Chabal, 2009; Pasternack et al., 2008).

Table 1 presents a list of references in which APTES was used to hopefully graft biomolecules on different substrates. The survey is exhaustive for stainless steeel substrates relevant for the field of biomaterials and illustrative for other substrates. Additional references are: El-Ghannam et al., 2004; Kim et al., 2010; Sasou et al., 2003; Sarath Babu et al., 2004; Quan et al., 2004 ; Subramanian et al., 1999 ; Jin et al., 2003 ; Cho & Ivanisevic, 2004 ; Katsikogianni & Missirlis, 2010 ; Sordel et al., 2007 ; Toworfe et al., 2006 ; Balasundaram et al., 2006 ; Doh & Irvine, 2006 ; Palestino et al., 2008 ; Son et al., 2011 ; Koh et al., 2006 ; Mosse et al., 2009 ; Weng et al., 2008 ; Charbonneau et al., 2011 ; Iucci et al., 2007 ; Chuang et al., 2006 ; Schuessele et al., 2009 ; Ma et al., 2007 ; Toworfe et al., 2009 ; Zile et al., 2011 ; Sargeant et al., 2008 ; Lapin & Chabal, 2009. Table 1 indicates the substrate and linker used, the main conditions of the APTES treatment and the evaluation of the surface treatment regarding biomolecule activity with the blank used for comparison. The table also presents the main data obtained by surface characterization. In some systems, no covalent grafting was aimed. In other systems, although it was aimed, there is no direct evidence for the formation of covalent bonds between the biomolecules and the substrate surface. On the other hand, the evaluation of the bio-efficacy was never based on comparisons involving a complete set of blanks: treatment with the biomolecule without silanization, without linker, without silanization and linker. In a study of surface modification with the aim to enhance mineralization, it has been demonstrated that APTES-coated glass retains a homopolymer with monoester phosphate groups, poly[(2-methacryloyloxy)ethyl phosphate], by proton transfer and electrostatic interaction, while the retention of a neutral homopolymer, poly[2-(acatoacetoxy)ethyl methacrylate], was attributed to covalent linkage by reductive amination between the keto groups of the polymer and the surface amine functions (Jasienak et al., 2009). The retention of the diblock copolymer seemed to occur via segments allowing covalent bonds to be formed. In Table 1, several systems show an improved behavior which may only be attributed to non covalent bonding between the biomolecule and the silanized substrate. In contradiction with frequent implicit considerations, the occurrence or improvement of bioactivity as a result of surface treatments does not demonstrate that the chemical schemes which motivated the treatments worked in reality. This question is crucial as many organic reactions that work well in solution are difficult to apply at solid surfaces (Kohli et al., 1998).

Silanization with APTES for Controlling the Interactions Between Stainless Steel and
Biocomponents: Reality vs Expectation

115

Stainless steels (SS) are extensively used in biomaterials researches and other applications involving contact with biologic compounds, owing to their adaptable mechanical properties, their manufacturability and their outstanding corrosion resistance. For instance, SS may be used in the manufacture of vascular stents, guide wires, or other orthopedic implants (Hanawa, 2002, Ratner, 2004). In these conditions, SS are subjected to the adsorption of biomolecules (proteins, polysaccharides, lipids) and biological materials (cellular debris). The surface modification of SS may thus be important to orient the host response as desired. The present work is dedicated to the surface composition of 316L SS surfaces at different stages of the procedure used to graft a protein via the use of APTES and of a bifunctional agent expected to link the NH_2-terminated silane with NH_2 groups of the protein. Glucose oxidase was chosen as a model protein for reasons of convenience owing to previous works related to microbiologically influenced corrosion (Dupont et al., 1998, Landoulsi et al., 2009, Landoulsi et al., 2008b, Landoulsi et al., 2008c). A particular attention is given to (i) the real state of the interface (composition, depth distribution of constituents) at different stages and (ii) the mode of protein retention. Therefore, X-ray photoelectron spectroscopy is used in a way (angle resolved measurements, reasoned peak decomposition, validation by quantitative relationships between spectral data) to provide a speciation in terms of classes of compounds (silane, protein, contaminants), using guidelines established in previous works (Genet et al., 2008; Rouxhet & Genet, 2011). Water contact angle measurements are used to address the issue of the presence of contaminants and the perspectives of avoiding it.

Substrate	Linker	Biomolecule	Reference
a. Substrate preparation			
b. APTES treatment			
c. Evaluation of efficiency regarding biomolecule activity. Substrate taken as blank			
d. Interface characterization			
Stainless steel	EDC	Alginate	Yoshioka et al., 2003
a. Sonication in acetone, 5 min; heating 2 h at 500°C in air.			
b. In toluene, 1h; rinsing in toluene and ethanol; sonication in ethanol, 5 min; drying in air; curing 10 min at 105°C.			
c. Preventing adsorption of blood-clotting proteins. Blanks = native, silanized.			
d. XPS: elemental concentration, consistent evolution according to reaction steps; C 1s peak, demonstration of alginate retention, majority of carbon of C-(C,H) type at all stages.			
Stainless steel	GA	Lysozyme	Minier et al., 2005
a. Acid etching at 60 °C; rinsing in water; drying under N_2 gas flow.			
b. In ethanol/water, 3 min; curing 1 h at 100-150 °C in air; rinsing with water.			
c. Increase of the enzymatic activity in bacterial lysis. Blanks = native + enzyme, silanized + enzyme.			
d. IRRAS: Characteristic bands of APTES. XPS: elemental concentration, consistent evolution according to reaction steps; N/Si ratio vs photoelectron collection angle, consistent evolution.			
Stainless steel stent	none	Chitosan/heparin LbL film	Meng et al., 2009
a. Cleaning in ethanol/water (1/1, v/v); rinsing in water; drying under reduced pressure, 24 h at 30 °C.			
b. In ethanol, 4 h at 37 °C; rinsing in water, drying in air at 50 °C.			
c. Promoting re-endothelialization after stent implantation, improvement of			

haemocompatibility (in vivo and in vitro tests). Blank = native stainless steel stent, native+chitosan.

d. QCM on model substrate slide: in situ monitoring of the LbL film growth on APTES-coated silicon substrate.

Stainless steel screw	GA	Fibrinogen + bisphosphonate	North et al., 2004

a. Sonication in acetone; acid etching; treatment in H_2O_2/NH_4OH solution, 5 min at 80 °C.

b. Vapor deposition, 10 min at 60 °C; curing 1 h at 150°C; sonication in xylene.

c. Improvement of fixation of screws in rat tibia. Blank = test of bisphosphonate action.

d. Ellipsometry on model substrate slide: consistent increase of the film thickness according to reaction steps.

Titanium	SMP	RGDC peptide	Xiao et al., 1997

a. Acid etching; rinsing with different solvents; outgassing.

b. In dry toluene, 120 °C, 3 h plus variants; sonication in various organic solvents and water.

c. -

d. Ellipsometry and radiolabeling: growth of silane surface layer upon repeating treatments. XPS peak shapes: semi-quantitatively consistent evolution according to reaction steps.

Titanium	GA	Chitosan	Martin et al., 2007

a. Polishing; cleaning in different solvents followed by nitric acid passivation or piranha treatment.

b. In toluene, 24 h; sonication in toluene, ethanol and water.

c. -

d. XPS: elemental concentration, consistent evolution according to reaction steps.

Titanium	none	Heparin/fibronectin	Li et al., 2011

a. Polishing; sonication in different solvents, drying 2 h at 60 °C ; NaOH treatment, 2h at 80°C.

b. In anhydrous ethanol, 10 h, sonication in ethanol; curing 6 h at 120 °C.

c. Improvement of blood compatibility and promotion of endothelialization. Blanks = native, silanized.

d. XPS : survey spectra consistent with treatments. AFM: small clumps after APTES step, and after APTES + protein step. Wet chemical analysis of proteins: consistent evolution according to reaction steps.

Ti-6Al-4V	SMP	Cyclic peptides	Porté-Durrieu et al., 2004

a. Substrate oxidation; outgassing at 150 °C.

b. In dry hexane under argon; rinsing with dry hexane; outgassing.

c. Increase of osteoprogenitor cells attachment. Blanks = native, native + peptide.

d. XPS: elemental concentration, consistent evolution according to reaction steps; N 1s peak, demonstration of peptide retention; C 1s peak, majority of carbon of C-(C,H) type at all stages.

Co-Cr-Mo; Ti-6Al-4V	GA	Trypsin	Puleo, 1997

a. Cleaning in different solvent, acid passivation.

b. In water 3 h or acetone 10 min; curing at 45 °C overnight.

c. Decrease of the loss of enzymatic activity as a function of time for Co-Cr-Mo, no effect

Silanization with APTES for Controlling the Interactions Between Stainless Steel and
Biocomponents: Reality vs Expectation

117

for Ti-6Al-4V. Blank = native + enzyme, silanized + enzyme.
d. Wet chemical analysis of amino groups on silanized substrate only.

Magnesium	Ascorbic acid	BSA	Killian et al., 2010

a. Surface polishing; rinsing in water and ethanol.
b. In toluene, 24 h at 70 °C; rinsing in different solvents; sonication in ethanol.
c. -
d. XPS: elemental concentration, consistent evolution according to reaction steps. Tof-SIMS: demonstration of silanization and protein retention.

Tantalum coating	DS3; DSS; DSC; CDI	Collagen	Müller et al., 2005

a. Acid etching; rinsing with different solvents; drying in vacuum.
b. In boiling toluene under argon; rinsing in chloroform, methanol, and in water.
c. Increase of cell adhesion and proliferation (stem cell culture and subcutaneous implantation). Blanks = native, silanized.
d. Wet chemical analysis: amino groups, collagen, consistent with expectation. AFM: typical structure of fibrillar collagen.

SiO_2, TiO_2, Si_3N_4	GA	Antigen	Kim et al., 2009b

a. Plasma treatment.
b. In ethanol/water (95/5, v/v), rinsing with ethanol ; curing 15 min at 120 °C.
c. Antigen/antibody test. Influence of plasma exposure time.
d. Water contact angle and XPS on cleaned substrate, influence of plasma exposure time.

Silicon dioxide	SMP	RGD peptide	Davis et al., 2002

a. Oxidation and hydroxylation; rinsing with different solvents.
b. In dry toluene, 120 °C, 3 h; sonication in various organic solvents and water.
c. Increase of fibroblast proliferation. Blanks = no RGD.
d. XPS: elemental concentration, consistent evolution according to reaction steps. AFM: small clumps after APTES step ; larger clumps after APTES+SMP step.

Silicon dioxide	GA	Glucose oxidase	Libertino et al., 2008

a. Oxidation in a solution of ammonia and hydrogen peroxide.
b. Vapor deposition; curing 40 min at 80 °C under vacuum.
c. Slight decrease of the enzyme activity when the surface is not oxidized (a). No blank.
d. AFM: consistent evolution of surface roughness according to reaction steps. XPS: elemental concentration, consistent evolution according to reaction steps.

Silicon dioxide	LC-SPDP	Tagged Kcoil peptides	Boucher et al., 2009

a. Piranha treatment, 10 min at 100°C; rinsing in water; drying in air; storing in vacuum.
b. In anhydrous toluene, 3 h; curing 1.5 h at 120°C; sonication in freshly distilled toluene.
c. Improvement of binding efficiency (amount, affinity) of Ecoil-tagged EGF. Blank = silanized + LC-SPDP with blocked termination.
d. Ellipsometry and water contact angle: consistent evolution according to reaction steps.

Glass	none	Proteins from ECM, RGD peptide	Siperko et al., 2006

a. Cleaning in different solvent.
b. In anhydrous ethanol + acetic acid, followed by addition of ultrapure water, 5 min; curing 15 min at 120°C.
c. Improvement of osteoblast adhesion and growth. Blank = silanized.

d. AFM: increase of surface roughness, presence of nanostructures presumably due to proteins or peptide.			
Poly(dimethylsiloxane)	SSMCC	DNA	Vaidya & Norton, 2004
a. Surface plasma oxidation. b. Vapor deposition, 30 min, 100 °C. c. Hybridization of attached DNA. No blank. d. XPS: elemental composition and peak shapes, demonstration of silane deposition.			
Polystyrene	BS³, GMBS	IgG	North et al., 2010
a. Plasma treatment. b. In ethanol+acetic acid; rinsing in methanol. c. - d. Fluorescence labeling: increase of amount of immobilized proteins by silanization + linker step.			
Cellulose	SMP	RGDC peptide	Bartouilh de Taillac et al., 2004
a. Drying, outgassing. b. In dry toluene under argon, 1.5 h; rinsing with dry toluene; outgassing. c. Increase of osteoprogenitor cells attachment and proliferation. Blank = native. d. XPS: qualitative consistency with expectations but probable Si contamination on final product.			

BSA = bovine serum albumin
BS³ = bis(sulfosuccinimidyl) suberate
CDI = 1,1'-carbonyldimidazole
DSC = N,N'-disuccinimidyl-carbonate
DSS = N,N'-disuccinimidyl-suberate
DS3 = N,N'-disulphosuccinimidyl-suberate
EDC = 1-ethyl-3-(3-dimethylaminopropyl)carbodiimide.
ECM = extracellular matrix
EGF = epidermal growth factor
GA = glutaraldehyde
GMBS = 4-maleimidobutyric acid N-hydrosuccinimide ester
LbL = layer-by-layer
LC-SPDP = Succinimydil 6-[30-(2-pyridildithio)-propionamido]hexanoate
QCM = quartz crystal microbalance
RGDC = arginine - glycine - aspartic acid – cysteine
SMP = N-succinimidyl-3-maleimidopropionate
SSMCC = sulfosuccinimidyl-4-(N-maleimidomethyl)-cyclohexane-1-carboxylate
TNBS = 2,4,6-trinitrobenzenesulfonic acid

Table 1. Illustration of the use of APTES for retaining biomolecules on surfaces. The meaning of a, b, c, and d is given in the upper box.

2. Materials and methods

2.1 Materials

Disks of 316L SS (0.74 cm², from Outokumpu Stainless AB, ARC) were used. The bulk chemical composition of 316L SS was given previously (Landoulsi et al., 2008a). All chemicals (NaCl, Na_2SO_4, $NaNO_3$ $NaHCO_3$, $NaHPO_4$, Na_2PO_4) were provided by Prolabo (VWR, France) and ensured 99% minimum purity. 3-aminopropyl(triethoxysilane) (APTES, 99%), glucose oxidase (Gox, EC 1.1.3.4, 47200 U.g⁻¹) from *Aspergillus Niger* and D-glucose

Silanization with APTES for Controlling the Interactions Between Stainless Steel and
Biocomponents: Reality vs Expectation

119

were purchased from Sigma-Aldrich (France). Bis(sulfosuccinimidyl) suberate (BS) was purchased from Pierce (Rockford, IL, USA).

2.2 Stainless steel surface preparation

The samples (both faces and perimeter) were polished with 1 μm diamond suspension (Struers, Denmark), rinsed in binary mixture of milliQ water/ethanol (1/1, v/v) in a sonication bath (70W, 40 kHz, Branson, USA) and dried under nitrogen gas flow.The samples were then immediately immersed for 48 h in synthetic aqueous medium (NaCl 0.46 mmol.L^{-1}, Na$_2$SO$_4$ 0.26 mmol.L^{-1}, NaNO$_3$ 0.2 mmol.L^{-1}, NaHCO$_3$ 3.15 mmol.L^{-1}, pH about 8), abundantly rinsed with milliQ water (Millipore, Molsheim, France) and dried under nitrogen gas flow. These samples are considered as native SS and designated as "nat".

2.3 Surface treatment procedure

The silanization of nat samples was performed in the gas phase with APTES. To this end, nat samples were placed, together with a small vial containing 100 μL APTES, in a 7 dm^3 closed recipient under vacuum for 30 min at room temperature. The samples were cured for 1 h at 100°C under vacuum, rinsed and incubated for 6 h in milliQ water to eliminate the excess of non attached silanes. These are called "sil".

Both nat and sil samples were subjected to treatments with the coupling agent (BS) and/or the enzyme (Gox) as detailed below:

i. "+BS" and "sil+BS" obtained after BS treatment (10 mM in milliQ water) for 30 min on nat and sil samples, respectively,

ii. "+Gox" and "sil+Gox" obtained after Gox treatment (0.1 mg.mL^{-1} in phosphate buffer pH~6.8) for 2 h, on nat and sil samples, respectively,

iii. "+BS+Gox" and "sil+BS+Gox" obtained after both BS and Gox treatment (according to the procedure described above) on nat and sil samples, respectively.

After BS or Gox treatment, the samples were rinsed three times with milliQ water and dried under nitrogen gas flow.

In order to clarify the issue of surface contamination, nat samples were further cleaned by UV-ozone treatment (UVO Cleaner, Jelight Co, Irvine, Ca, USA) during 20 minutes. They were then placed in a Petri dish, left in the laboratory environment, and submitted to water contact angle measurements as a function of time.

2.4 X-ray photoelectron spectroscopy

XPS analyses were performed using a Kratos Axis Ultra spectrometer (Kratos Analytical, UK), equipped with a monochromatized aluminum X-ray source (powered at 10 mA and 15 kV) and an eight channeltrons detector. No charge stabilization device was used on these conductive samples. Analyses were performed in the hybrid lens mode with the slot aperture; the resulting analyzed area was 700 μm × 300 μm. A pass energy of 40 eV was used for narrow scans. In these conditions, the full width at half maximum (FWHM) of the Ag 3d$_{5/2}$ peak of clean silver reference sample was about 0.9 eV. The samples were fixed on the support using a double-sided adhesive conducting tape. The pressure in the analysis chamber was around 10^{-6} Pa. The photoelectron collection angle θ between the normal to the sample surface and the electrostatic lens axis was 0° or 60°. The following sequence of spectra was recorded: survey spectrum, C 1s, O 1s, N 1s, P 2p, Cr 2p, Fe 2p, Ni 2p, Mo 3d, Na 1s, S 2p, Si 2p and C 1s again to check for charge stability as a function of time, and absence of sample degradation. The binding energy scale was set by fixing the C 1s component due to carbon only bound to carbon and hydrogen at 284.8 eV. The data treatment was performed with the Casa XPS software (Casa Software Ltd., UK). The peaks were decomposed using a linear baseline, and a component shape defined by the product of a Gauss and Lorentz function, in the 70:30 ratio, respectively. Molar concentration ratios were calculated using peak areas normalized according to the acquisition parameters and the relative sensitivity factors and transmission functions provided by the manufacturer.

2.5 Atomic force microscopy (AFM)

The surface topography was examined using a commercial AFM (NanoScope III MultiMode AFM, Veeco Metrology LLC, Santa Barbara, CA) equipped with a 125 μm $\times$ 125 μm $\times$ 5 μm scanner (J-scanner). A quartz fluid cell was used without the O-ring. Topographic images were recorded in contact mode using oxide-sharpened microfabricated Si_3N_4 cantilevers (Microlevers, Veeco Metrology LLC, Santa Barbara, CA) with a spring constant of 0.01 N.m^{-1} (manufacturer specified), with a minimal applied force (<500 pN) and at a scan rate of 5-6 Hz. The curvature radius of silicon nitride tips was about 20 nm. Images were obtained at room temperature (21-24°C) in milliQ water. All images shown in this paper were flattened data using a third order polynomial. The surface roughness (R_{rms}) was computed over an area of 1 μm $\times$ 1 μm using the Veeco software.

2.6 Water contact angle measurements

Water contact angles were measured at room temperature using the sessile drop method and image analysis of the drop profile. The instrument, using a CCD camera and an image analysis processor, was purchased from Electronisch Ontwerpbureau De Boer (The Netherlands). The water (milliQ) droplet volume was 0.3 μL, and the contact angle was measured 5 s after the drop was deposited on the sample. For each sample, the reported value is the average of the results obtained on 5 droplets.

3. Results

AFM images obtained in water on SS samples after different treatments are presented in Figure 1. The nat sample showed the presence of nanoparticles with different sizes (nat, Figure 1, R_{rms} = 3.2 nm), in agreement with previous results. The formation of nanoparticles, presumably made of ferric hydroxide, resulted from oxidation occurring during the 48 h of immersion subsequent to polishing (Landoulsi et al., 2008a). The surface of silanized SS exhibited particles with a bigger size (sil, Figure 1) and the roughness decreased slightly (sil, Figure 1, R_{rms} = 2.4 nm). The treatment with Gox, with or without previous treatment with BS, led to the formation of particles with a more uniform size and a higher density, in comparison with nat sample, and no appreciable change of surface roughness (R_{rms} = 1.7 nm for sil+Gox and 2.5 nm for sil+BS+Gox, Figure 1).

XPS is a suitable method to obtain information regarding the different constituents at the surface (substrate, silane, other organic compounds). The elemental composition of the samples is given in Table 2. Representative C 1s and O 1s peaks recorded on SS surface prior to and after silanization, and after further treatments are given in Figure 2. After BS and/or Gox treatment of silanized SS, a relative increase of the component around 531.2 eV in the O 1s peak was observed (Figure 2). In the C 1s peak, an increase of the components at 286.3 and 288.7 eV was also clear, while the main component remained at 284.8 eV. The same

tendency regarding the C 1s and O 1s peaks was also observed after BS and/or Gox
treatment of nat sample, but was less pronounced (data not shown). It may be attempted to

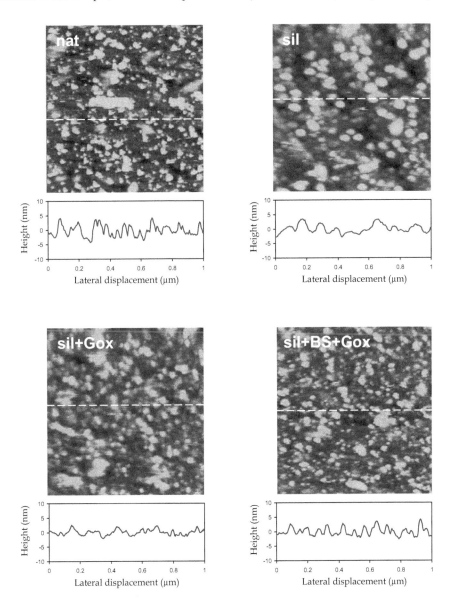

Fig. 1. AFM height images (1×1 µm^2, contact mode, in water; z scale 10 nm) of native (nat),
silanized stainless steel (sil), the same after adsorption of glucose oxidase (sil+Gox) and after
treatment with glucose oxidase subsequent to treatment with the coupling agent
(sil+BS+Gox). Cross sections were taken at the place indicated by the dashed lines.

extract chemical information by careful decomposition of the peaks. This requires to impose reasonable constraints (number of components, full width at half maximum FWHM) in order to insure reliable comparisons, and to check the chemical relevance of the results by examining correlations between spectral data of different natures (Genet et al., 2008; Rouxhet & Genet, 2011). In previous studies (Landoulsi et al., 2008a; Landoulsi et al., 2008b), we have demonstrated the usefulness of this approach, even when the evolution of the C 1s and O 1s peak shape is weak, in order to obtain information on the amount and the nature of organic and inorganic constituents on SS surfaces.

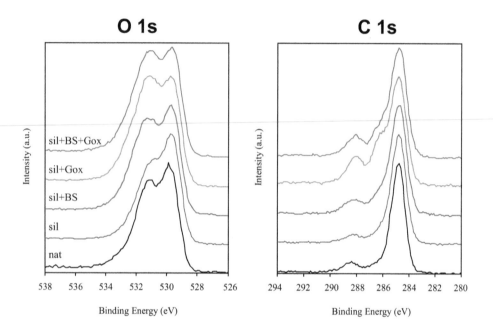

Fig. 2. O 1s and C 1s peaks of native (nat), of silanized stainless steel (sil), of the same after treatment with the coupling agent (sil+BS), after adsorption of glucose oxidase (sil+Gox) and after treatment with glucose oxidase subsequent to treatment with the coupling agent (sil+BS+Gox).

Figure 3 presents typical O 1s, N 1s and C 1s XPS peaks recorded on native SS (nat), silanized (sil) and the same after Gox treatment (sil+Gox). For the decomposition of these peaks, reasonable constraints were applied, based on our experience with the XPS analysis of biosurfaces (Genet et al., 2008, Rouxhet & Genet, 2011). The C 1s peak was decomposed in four components, the FWHM of which were imposed to be equal: (i) a component at 284.8 eV due to carbon only bound to carbon and/or hydrogen [C-(C,H)]; (ii) a component at about 286.3 eV due to carbon making a single bond with oxygen and/or nitrogen [C-(O,N)] in alcohol, amine, or amide; (iii) a component at 287.8 eV due to carbon making one double bond or two single bonds with oxygen (C=O, O-C-O) and (iv) a component at 288.7 eV attributed to carboxyl or ester functions [(C=O)-O-R].

Silanization with APTES for Controlling the Interactions Between Stainless Steel and
Biocomponents: Reality vs Expectation

123

The O 1s peak was decomposed in three components (Landoulsi et al., 2008a). The first one, at 529.7 eV, is due to inorganic oxygen in metal oxides (M–O) (NIST Database). The FWHM of the two other components were arbitrarily imposed to be equal. The component at about 531.2 eV may be due to oxygen making a double bond with carbon (C=O including amide and carboxyl group) and to oxygen of carboxylate. Contributions of metal hydroxides (M–O–H) as well as oxygen bound to silicon [Si–O] in silane are overlapping with this component (NIST Database; Genet et al., 2008). The last component, at 533.1 eV, is attributed to oxygen making single bonds with carbon (C–O–H of alcohol and carboxyl, C–O–C of ether and ester).

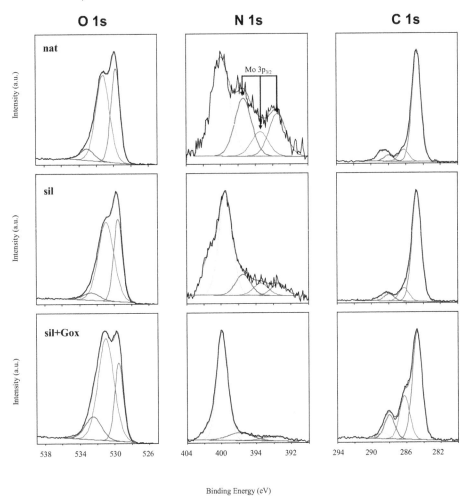

Fig. 3. Decomposition of O 1s, N 1s and C 1s peaks recorded on: native stainless steel (nat), silanized stainless steel (sil) and the same after adsorption of glucose oxidase (sil+Gox). The N 1s peak is overlapped with a Mo $3p_{3/2}$ contribution.

The N 1s peak showed a main component at 400.0 eV attributed to amide or amine (N–C). An additional component appeared clearly near 401.6 eV in silanized SS samples (sil and sil+Gox, Figure 3), indicating the presence of protonated amines. The contributions at lower binding energies in the N 1s spectral window are due to Mo $3p_{3/2}$ components (Olefjord & Wegrelius, 1996); the reliability of their quantification was checked by comparison with the Mo concentration deduced from the Mo $3p_{3/2}$ and Mo 3d peaks (Landoulsi et al., 2008a). The N 1s contribution of sil shows a shape which is in agreement with spectra reported in the literature for APTES-modified surfaces (Suzuki et al., 2006; Xiao et al., 1997). It was not found justified to decompose it in three components attributed to amine, amide and protonated amine, respectively, as done in (Suzuki et al., 2006). The surface concentrations (mole fraction) associated with the components of C 1s, O 1s and N 1s peaks are given in Table 2.

	Ni^{ox}	Ni^{met}	Ni$_{tot}$	Fe^{ox}	Fe^{met}	Fe$_{tot}$	Cr^{ox}	Cr^{met}	Cr$_{tot}$	Mo^{ox}	Mo^{met}	Mo$_{tot}$	Si^0
nat	1.23	0.73	**1.95**	7.59	1.57	**9.16**	4.69	0.33	**5.02**	0.34	0.12	**0.46**	0.38
+BS	0.81	0.54	**1.35**	8.32	1.89	**10.21**	5.96	0.40	**6.36**	0.46	0.12	**0.57**	0.29
+Gox	0.59	0.51	**1.10**	8.63	1.70	**10.33**	5.53	0.33	**5.86**	0.37	0.12	**0.49**	-
+BS+Gox	0.66	0.39	**1.05**	8.90	1.43	**10.33**	4.56	0.30	**4.87**	0.32	0.12	**0.44**	0.36
sil	0.47	0.33	**0.80**	7.89	1.22	**9.10**	4.59	0.27	**4.86**	0.31	0.10	**0.40**	0.37
sil+BS	0.47	0.43	**0.91**	6.91	1.45	**8.36**	5.24	0.33	**5.57**	0.35	0.11	**0.45**	0.32
sil+Gox	0.60	0.38	**0.98**	6.14	1.29	**7.42**	4.95	0.29	**5.23**	0.32	0.12	**0.44**	0.26
sil+BS+Gox	0.54	0.32	**0.86**	6.22	1.16	**7.39**	4.58	0.24	**4.82**	0.34	0.09	**0.42**	0.30

	C$_{288.7}$	C$_{287.8}$	C$_{286.3}$	C$_{284.8}$	C$_{tot}$	O$_{533.1}$	O$_{531.2}$	O$_{529.7}$	O$_{tot}$	O$_{org}$	N$_{401.6}$	N$_{400}$	N$_{tot}$	Si$_{org}$	Σorg
nat	2.45	2.29	3.92	39.62	**48.28**	2.36	19.02	12.60	**33.98**	7.92	0.00	0.73	**0.73**	0.44	57.37
+BS	2.89	2.08	3.88	34.87	**43.71**	2.48	18.96	13.86	**35.30**	7.64	0.00	1.21	**1.21**	0.43	52.99
+Gox	2.03	2.16	4.78	35.56	**44.53**	2.97	20.96	11.27	**35.20**	7.50	0.13	1.34	**1.48**	-	53.50
+BS+Gox	2.55	2.60	5.89	30.02	**41.06**	3.96	22.01	12.57	**38.54**	9.46	0.23	1.35	**1.58**	0.89	52.99
sil	1.15	2.37	4.57	36.13	**44.23**	1.95	21.22	12.28	**35.45**	5.93	0.63	1.54	**2.17**	2.29	54.61
sil+BS	2.03	3.30	6.48	32.72	**44.53**	1.07	23.16	9.71	**33.94**	8.49	0.66	2.65	**3.31**	2.13	58.47
sil+Gox	1.18	5.97	10.39	25.08	**42.62**	4.74	21.09	8.86	**34.70**	11.78	0.59	5.16	**5.75**	1.74	61.90
sil+BS+Gox	1.46	4.89	8.31	30.39	**45.05**	2.87	22.24	8.52	**33.63**	9.86	0.50	4.30	**4.80**	2.01	61.72

Table 2. Surface concentration (mole fraction (%) computed over the sum of all elements except hydrogen) of elements determined by XPS ($\theta = 0°$) on stainless steel samples.

The Si 2p peak was decomposed in two components, at 99.3 and 101.8 eV, attributed to non oxidized silicon in SS (Si^0) and to silicon of silane (Si$_{org}$), respectively. It was not decomposed in Si $2p_{3/2}$ and Si $2p_{1/2}$ contributions because these are very close in energy. The decomposition procedure for Fe 2p$_{3/2}$, Cr 2p, Ni 2p$_{3/2}$, and Mo 3d peaks was described before (Landoulsi et al., 2008a). The Fe concentration may be underestimated due to the procedure used to treat the complex baseline of the Fe 2p peak. The distinction between contributions of oxidized (M^{ox}) and nonoxidized (M^{met}) metal elements was easily made. The concentrations obtained are also given in Table 2.

The concentration of oxygen present in organic compounds O_{org} may not directly be deduced from the O 1s peak owing to the overlap with inorganic hydroxide. However, for biological systems the sum of O and N concentrations may be evaluated by the

Silanization with APTES for Controlling the Interactions Between Stainless Steel and
Biocomponents: Reality vs Expectation

125

concentration of carbon in oxidized form, C_{ox}, in consistency with alcohol, primary amine, primary amide and ester functions. Accordingly,

$$O_{org} = C_{ox} - N = C_{286.3} + C_{287.8} + C_{288.7} - N_{tot} \tag{1}$$

where the name of an element in italic designates its concentration and the number in subscript designates the binding energy of the peak component. Errors would occur in case of a high concentration of polysaccharides ($C_{ox}/O = 6/5$) or carboxyl ($C_{ox}/O = 1/2$) (Genet et al., 2008 ; Landoulsi et al., 2008a).
The sum of the concentrations of the elements present in organic compounds is then given by:

$$\sum org = C_{tot} + O_{org} + N_{tot} + Si_{org} = C_{tot} + C_{ox} + Si_{org} \tag{2}$$

For sake of uniformity, all spectral data involved in correlations below are ratioed to $\sum org$ (Table 2).
The concentration of the main elements or functions due to organic compounds, obtained at photoelectron collection angle $\theta = 0°$, is plotted in Figure 4 as a function of the same quantity obtained at $\theta = 60°$. A 1:1 relationship is obtained for all elements or functions and all samples, indicating no significant effect of the photoelectron collection angle θ on the relative contribution of the constituents of the organic adlayer.

4. Discussion

4.1 Passive film composition

Table 2 shows that the apparent concentrations of metal elements varied only slightly according to the surface treatment. The main change concerned the decrease of the Fe^{ox} concentration for sil+BS, sil+Gox and sil+BS+Gox samples. However, no change in the shape of the Fe $2p_{3/2}$ peak was observed (data not shown). For these samples, a significant decrease of the molar concentration of $O_{529.7}$ was also noticed (Table 2). It appears that the oxide layer of SS passive film, after incubation in the aqueous medium for 48 h (nat sample), was mainly constituted with a mixture of Fe and Cr oxides/hydroxides and small amounts of partially oxidized Ni and Mo. This is in agreement with a previous study (Landoulsi et al., 2008a), however in the latter, the stoichiometry of the passive film was not computed.
The O 1s component at 529.7 eV is due to metal oxides. By considering that Mo^{ox} is in the form of MoO_3 (Landoulsi et al., 2008a), the difference between the $O_{529.7}$ concentration and three times the Mo^{ox} concentration should be due to Fe and Cr oxides. Figure 5 presents the relation between this difference and the sum of Fe^{ox} and Cr^{ox} concentrations. All data show reasonable linear regressions. The shift of the dots along the line when the photoelectron collection angle changes from 0° to 60° is due to the presence of the organic constituents on top of stainless steel. The average ratio between the y and x scales is 0.80 and 0.96 at $\theta = 0$ and 60°, respectively; the slope of the regression lines is 1.03 (s.d. 0.23) and 1.42 (s.d. 0.12), respectively. Thus, the ratio oxide/metal ions in chromium and iron oxyhydroxides is of the order of 1 to 1.5.
The evaluation of the quantity of hydroxide associated to Fe and Cr is complex due to the multiple chemical functions overlapping in the $O_{531.2}$ component. Ni^{ox} is in the form of $Ni(OH)_2$ (component at ~855.6 eV (Briggs & Seah, 1990, Zhou et al., 2006), spectra not shown). The amount of oxygen associated to silane depends on the products of APTES

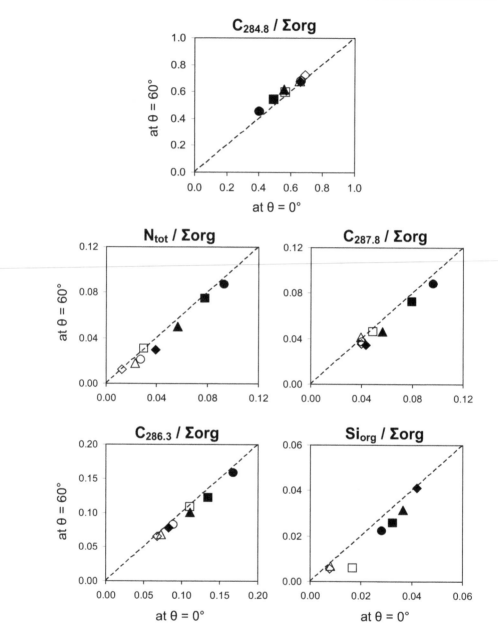

Fig. 4. Plots of molar concentrations ratioed to the sum of organic elements (Σorg) measured by XPS at $\theta = 60°$ *vs* $\theta = 0°$ on native (open symbols) or silanized stainless steel (closed symbols), as such ($\blacklozenge,\lozenge$) or further treated with coupling agent BS ($\blacktriangle,\triangle$), glucose oxidase ($\bullet,\circ$) or coupling agent followed by glucose oxidase ($\blacksquare,\square$). The dashed lines represent a y/x ratio of 1:1.

Silanization with APTES for Controlling the Interactions Between Stainless Steel and
Biocomponents: Reality vs Expectation

127

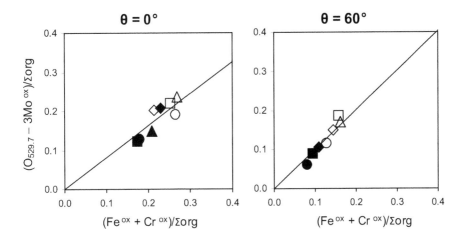

Fig. 5. Relations between molar concentrations ratioed to the sum of organic elements (Σorg) measured by XPS at θ = 0° (data from Table 2) and θ = 60° on native (open symbols) or silanized stainless steel (closed symbols), as such (◆,◇) or further treated with coupling agent BS (▲,△), glucose oxidase (●,○) or coupling agent followed by glucose oxidase (■,□).

reaction. The different possibilities, corresponding to the relative importance of grafting with respect to polymerization, are shown in Figure 6 and characterized by a, defined as the sum, over oxygen atoms which are not bound to a metal element, of the inverse of the number of bonds oxygen forms with silicon. Accordingly the concentration of inorganic hydroxide in the passive film may be given as follows:

$$OH_{inorg} = O_{tot} - O_{org} - O_{529.7} - a \times Si_{org} \qquad (3)$$

where a can take the values of 0, 0.5, 1, 1.5 or 2 (Figure 6).

Figure 7 shows the plot of the concentration of hydroxide which should be associated with Fe and Cr, considering different values of a. Taking silicon into account (a≠0) brings the silanized substrates better in line with the non-silanized substrates and the correlation improves as a increases, indicating that silane is polymerized and not just grafted. Depending on the value of a, the y : x ratio varies from 1.1 (a = 0) to 0.8 (a = 2).

These observations indicate that the stoichiometry of the Cr and Fe oxyhydroxide at the surface is close to (Fe,Cr)OOH. Many studies have reported a stratification in the passive film and the presence of Fe and Cr oxyhydroxides in the outermost layer (Le Bozec et al., 2001). In our case, no significant effect of the photoelectron collection angle appears in the $OH_{inorg}/O_{529.7}$ ratio (Figure 8), whatever the value selected for a to evaluate OH_{inorg}. However it must be kept in mind that the surface roughness revealed by Figure 1, which is of the order of the inelastic mean free path of photoelectrons in oxides, may mask the effect of a stratification.

$$H_2N - (CH_2)_3 - Si - \begin{bmatrix} O - M \dots \\ O - M \dots \\ O - M \dots \end{bmatrix} a = 0 \qquad\qquad H_2N - (CH_2)_3 - Si - \begin{bmatrix} O - M \dots \\ O - M \dots \\ O - Si \dots \end{bmatrix} a = 0.5$$

$$H_2N - (CH_2)_3 - Si - \begin{bmatrix} O - M \dots \\ O - Si \dots \\ O - Si \dots \end{bmatrix} a = 1 \qquad\qquad H_2N - (CH_2)_3 - Si - \begin{bmatrix} O - Si \dots \\ O - Si \dots \\ O - Si \dots \end{bmatrix} a = 1.5$$

$$H_2N - (CH_2)_3 - Si - \begin{bmatrix} O - M \dots \\ O - Si \dots \\ OH \end{bmatrix} a = 1.5 \qquad\qquad H_2N - (CH_2)_3 - Si - \begin{bmatrix} O - Si \dots \\ O - Si \dots \\ OH \end{bmatrix} a = 2$$

Fig. 6. Possible products of APTES reaction. "M" designates metal elements of stainless steel.

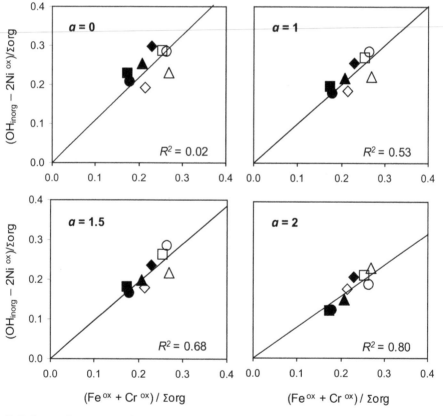

Fig. 7. Relations between molar concentrations ratioed to the sum of organic elements (Σorg) measured by XPS at $\theta = 0°$ (data from Table 2) on native (open symbols) or silanized stainless steel (closed symbols), as such ($\blacklozenge,\diamondsuit$) or further treated with coupling agent BS ($\blacktriangle,\triangle$), glucose oxidase ($\bullet,\circ$) or coupling agent followed by glucose oxidase ($\blacksquare,\square$).

Silanization with APTES for Controlling the Interactions Between Stainless Steel and
Biocomponents: Reality vs Expectation

129

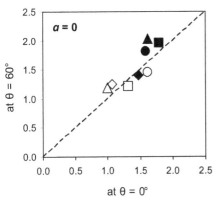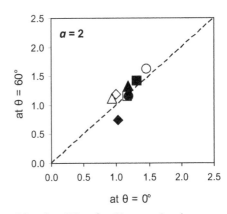

Fig. 8. Plots of OH$_{inorg}$ / O$_{529.7}$ ratio measured by XPS at $\theta = 60°$ vs $\theta = 0°$ on native (open symbols) or silanized stainless steel (closed symbols), as such ($\blacklozenge,\lozenge$) or further treated with coupling agent BS ($\blacktriangle,\triangle$), glucose oxidase ($\bullet,\bigcirc$) or coupling agent followed by glucose oxidase ($\blacksquare,\square$). The dashed lines represent a y/x ratio of 1:1.

4.2 Chemical speciation of the organic adlayer

Table 2 reveals an increase of N_{tot} concentration as a result of surface treatments, and an increase of Si_{org} concentration for samples prepared with APTES treatment. However, the concentration of carbon is high and remains almost unchanged, suggesting that organic contaminants, mainly hydrocarbon-like compounds, are always dominating in the organic adlayer. If nitrogen was exclusively due to amide functions (N-C=O) as in the peptide link of proteins, and if the C 1s component at 287.8 eV was exclusivley due to amide, a 1:1 correlation would be found between the concentrations of $C_{287.8}$ and N_{tot}. This is indeed observed (Figure 9a) for the set of samples involving the silanized substrate. As nitrogen is partly in the form of silane, relevant alternatives for the abscissa scale may be the concentration of N_{400} or the difference between the concentrations of N_{tot} and Si_{org}. If polysaccharides were present with protein, the $C_{287.8}$ concentration should be corrected by subtracting the contribution of acetal and thus replaced by $[C_{287.8} - (C_{286.3} - N_{400})/5]$ (Ahimou et al., 2007, Landoulsi et al., 2008a) or $[C_{287.8} - (C_{286.3} - N_{tot} + Si_{org})/5]$. A comparison between different plots in Figure 9 shows that the dots representative of samples prepared with non-silanized substrate remain clustered. The shift of the cluster along the ordinate scale according to the plot indicates that $C_{287.8}$ concentration is higher than what can be attributed to amide. On the other hand, the samples prepared with silanized substrate preserve a unit slope whatever the plot, with much higher values of the coordinates for samples exposed to the enzyme, with or without the linker. This reveals an excellent agreement between the increases of concentrations of nitrogen and of carbon attributed to peptidic links (N−C=O), which result from the Gox treatment. It also validates the C 1s peak decomposition and component attribution.

The meaning of the surface composition appears more clearly if it is summarized in terms of concentration of model molecular compounds. This approach was already used for microbial surfaces (Dufrêne & Rouxhet, 1996; Tesson et al., 2009), for food products (Rouxhet et al., 2008) and for stainless steel aged in different conditions (Landoulsi et al.,

2008b), considering proteins (Pr), polysaccharides (PS), and hydrocarbon-like compounds (HC) which represent mainly lipids in biological systems. The novelty here is to take silane (Sl) into account in addition to the three classes of biochemical compounds; the chemical composition and density considered for these model compounds are listed in Table 3. Accordingly, the proportion of carbon atoms due to each model compound X (C_X/C_{tot}) can be computed by solving the following system of equations:

$$(N_{tot} - Si_{org})/C_{tot} = 0.273 \times (C_{Pr}/C_{tot}) \tag{4}$$

$$O_{org}/C_{tot} = 0.312 \times (C_{Pr}/C_{tot}) + 0.833 \times (C_{PS}/C_{tot}) \tag{5}$$

$$Si_{org}/C_{tot} = 0.333 \times (C_{Sl}/C_{tot}) \tag{6}$$

$$1 = (C_{Pr}/C_{tot}) + (C_{PS}/C_{tot}) + (C_{Sl}/C_{tot}) + (C_{HC}/C_{tot}) \tag{7}$$

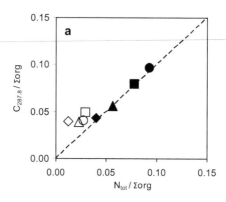

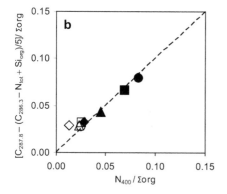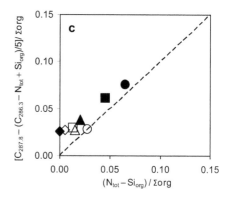

Fig. 9. Relations between molar concentrations ratioed to the sum of organic elements (Σorg) measured by XPS at θ = 0° (data from Table 2) on native (open symbols) or silanized stainless steel (closed symbols), as such (◆,◇) or further treated with coupling agent BS (▲,△), glucose oxidase (●,○) or coupling agent followed by glucose oxidase (■,□). The dashed lines represent a y/x ratio of 1:1.

Silanization with APTES for Controlling the Interactions Between Stainless Steel and
Biocomponents: Reality vs Expectation

131

The experimental concentration ratios (C_X/C_{tot}) can be converted into weight percentages of model compounds (Genet et al., 2008, Rouxhet & Genet, 2011), using the carbon concentration specific to each compound (Table 3).

	O/C	N/C	Carbon concent. (mmol.g⁻¹)	Density (g.cm⁻³)
Proteins	0.312*	0.273*	43.5	1.4
Silane	0.5	0.333	27.3	0.9
Polysaccharides	0.833	-	37.0	1.5
Hydrocarbons	-	-	71.4	0.9

* Data computed from the amino acid sequence of glucose oxidase on the basis of ProtParam tool available on the ExPASy molecular biology server (http://us.expasy.org).

Table 3. Chemical composition of organic compounds (molar concentration ratio of elements and concentration of carbon) and their densities.

Results, presented in Table 4, show the often dominating presence of hydrocarbon-like compounds and polysaccharides, for all samples. Both compounds are due to adventitious contamination which may originate from adsorption from air, as always observed for high surface energy solids (Caillou et al., 2008; Mantel et al., 1995), but also from aqueous media (Landoulsi et al., 2008a). They may be taken as a global way to reflect the amount of compounds which contain hydrocarbon chains and oxygen, such as esters, but excluding proteins and silane. The concentration of silane deduced for non-silanized samples is non negligible but highly variable, and may also be attributed to contamination. A drastic increase of silane concentration is observed for all silanized samples (Table 4).

	(wt. %)				Thickness (nm)	
	Pr	Sl	PS	HC	at θ = 0°	at θ = 60°
nat	2.9	5.8	29.5	61.7	3.6	2.2
+BS	(8.5)	6.1	28.3	57.1	3.2	2.1
+Gox	16.3	0.0	24.7	59.0	3.2	2.3
+BS+Gox	(7.3)	12.3	35.5	44.9	3.3	2.2
sil	0.0	29.0	22.2	48.8	3.8	2.7
sil+BS	(10.8)	25.5	25.2	38.5	3.9	2.8
sil+Gox	35.2	20.0	24.3	20.5	3.9	2.9
sil+BS+Gox	24.4	22.9	22.5	30.2	4.1	2.8

* The data between brackets are protein equivalent of BS products and have no physical meaning.

Table 4. Chemical composition (weight %) of the organic adlayer present on stainless steel samples, as deduced from XPS data and expressed in terms of classes of molecular compounds*. Thickness of the organic adlayer deduced from measurements at photoelectron collection angle θ = 0° and 60°.

The Gox treatment leads to a marked increase of the protein concentration on nat sample and a much stronger increase on sil sample. This may be attributed to physical adsorption and to the fact that protonated amine of silane favors adsorption by electrostatic attraction (Jasienak et al., 2009). The pH of the Gox solution (6.8) is indeed higher than the isoelectric point of glucose oxidase (4.9).

In the above computation, the concentration of the BS coupling agent could not be evaluated. The reaction of BS with NH_2 transforms an amine function into amide. If only one end of BS reacts with silane, the N_{400} concentration should be doubled, which is consistent with the increase found in Table 2. However, converting it into protein-equivalent gives a number with no physical meaning. When the second end of the coupling agent reacts with the protein, no additional nitrogen is incorporated, the evaluation of the protein concentration is correct but the suberate $(CH_2)_6$ chain is counted in the HC concentration. This has no important impact owing to the high concentration of hydrocarbon-like compounds. Despite the limitations regarding the accuracy of the data in Table 4, it is clear that prior silanization increases markedly the concentration of glucose oxidase (sil+Gox and sil+BS+Gox compared to nat+Gox); however the treatment with the coupling agent does not increase the amount of immobilized enzyme (sil+BS+Gox compared to sil+Gox).

4.3 State of stainless steel surface

It appears in Table 4 that the SS surface prior to and after silanization or enzyme immobilization is bearing a high amount of organic contaminants. It may be argued that this is due to improper cleaning protocols, inappropriate sample manipulation or contamination in the XPS spectrometer. Actually, a clean stainless steel surface is getting quickly contaminated in contact with the surrounding atmosphere, as revealed by water contact angles, which can be measured quickly in the same environment. A nat sample showed a water contact angle of 44° which increased to about 60° within a delay of a few hours (Figure 10). When a nat sample was further treated with UV-ozone to oxidize organic compounds, the water contact angle was lowered down to 12°. However it increased rapidly in contact with the surrounding atmosphere (Figure 10) to reach values above 40° in a few hours. Similar results were obtained with 304L stainless steel. Wet cleaning essentially standardizes the surface contamination; further cleaning leaves a material with a high surface energy, which adsorbs quickly significant amounts of contaminants (Caillou et al., 2008).

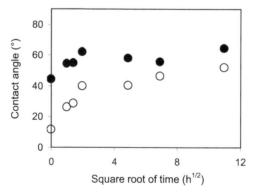

Fig. 10. Contact angle measurements as a function of incubation time in ambient atmosphere performed on (●) native stainless steel and (○) after UV-ozone treatment.

Silanization with APTES for Controlling the Interactions Between Stainless Steel and
Biocomponents: Reality vs Expectation

133

It must be emphasized that outgassing a material of high surface energy is not suitable to prevent adventitious contamination of the surface. Materials cleaned with UV-ozone treatment and showing a water contact angle below 5° reached an appreciable contact angle (20° for silica, 40° for stainless steel and gold) after a stay of 5 minutes in the vacuum of the spectrometer chamber. The high rate of contamination may be due to evacuation itself, owing to the increased proportion of organic compounds in the residual gas and to their increased rate of transfer (Caillou et al., 2008).

4.4. Thickness of the organic adlayer
No variation of the relative concentrations of organic compounds was revealed by angle-resolved XPS analysis (Figure 4) but an effect of stratification may be masked by the surface roughness. The evaluation of the adlayer thickness may clarify whether the increase of the silane or protein concentration resulting from the respective treatments reflects an addition of or a substitution by new compounds in the adlayer.

If the surface is considered as atomically smooth and the organic layer is continuous with a constant thickness, the apparent concentration ratio $[C]/[Cr]$ may be computed using the following equation:

$$\frac{[C]}{[Cr]} = \frac{i_{Cr}}{i_C}\frac{\sigma_C}{\sigma_{Cr}}\frac{\lambda_C^{Org}C_C^{Org}\left[1-\exp\left(\dfrac{-t}{\lambda_C^{Org}\cos\theta}\right)\right]}{\lambda_{Cr}^{Ox}C_{Cr}^{Ox}\exp\left(\dfrac{-t}{\lambda_{Cr}^{Org}\cos\theta}\right)} \tag{8}$$

i_C and i_{Cr} are the relative sensitivity factors of C and Cr, respectively, provided by the spectrometer manufacturer. The photoionization cross sections σ are 1 for C 1s and 11.7 for Cr 2p (Scofield, 1976). The superscripts Org and Ox designate the organic adlayer and the passive oxide layer, respectively. The concentration of Cr in the oxide layer (C_{Cr}^{Org}; between 0.015 and 0.020 mol.cm^{-3}, depending on the sample), was determined on the basis of the above discussion indicating that the oxide layer is constituted with FeOOH, CrOOH, Ni(OH)$_2$ and MoO$_3$ (*section 4.1*). Note that this is in agreement with a concentration of inorganic oxygen close to ($O_{tot} - O_{org}$), owing to the low concentration of silane. The concentration of carbon in the organic adlayer was determined on the basis of the surface composition modeled as detailed above, using the densities given in Table 3. The density of silane, was taken as the average between the densities of 3-aminopropyl(trimethylsilane) (0.8 g.cm^{-3}) and 3-aminopropyl(trimethoxysilane) (1.0 g.cm^{-3}). The electron inelastic mean free paths (IMFP) were calculated using the Quases program (http://www.quases.com) and the TPP2M formula (Tanuma et al., 1997), considering the matrix composition deduced above. For λ_{Cr}^{Ox}, values of 2.04 and 1.99 nm were computed for FeOOH and CrOOH, respectively (energy gap 2.3 and 1.6 eV, respectively). Considering an energy gap of 6 eV for the organic compounds, respective values of λ_C^{Org} and λ_{Cr}^{Org} were 3.79 and 3.04 nm in hydrocarbon-like compounds [(CH$_2$)$_n$], 3.59 nm and 2.89 nm in protein, 3.67 and 2.00 nm in polysaccharides [(C$_6$(H$_2$O)$_5$)$_n$], and 3.68 and 2.95 nm in silane [H$_2$N(CH$_2$)$_3$SiO$_{1.5}$].

The thickness of the organic adlayer deduced for photoelectron collection angles $\theta = 0°$ and 60° is given in Table 4. The difference between the values computed for the two photoelectron collection angles is due to the approximation of a smooth surface while the analyzed surface is rough at the scale of the inelastic mean free paths. The real organic adlayer thickness should be between the values given in Table 4. Comparison between nat and sil samples suggests that silane just adds up to the contaminants. A 3.0 nm thick adlayer containing 25 wt.% silane corresponds to 4.5 molecules.nm^{-2}. This value is consistent with a monolayer of silane, however the retained silane is mixed with a much larger amount of contaminants. The protein treatment of silanized substrates (compare sil+Gox and sil+BS+Gox to sil) led to a significant decrease of the amount of hydrocarbon-like compounds, while the adlayer thickness did not change appreciably. This suggests that the protein adsorption caused the displacement of part of contamination present on the silanized stainless steel surface in the form of hydrocarbon-like compounds.

5. Conclusion

The decomposition of the C 1s and O 1s peak provided a distinct evaluation of oxygen present in inorganic oxide, inorganic hydroxide and organic compounds. This led to a stoichiometry close to (Fe, Cr)OOH for iron and chromium species in the passive layer. The elemental composition of the organic adlayer was converted into concentration of model organic compounds: protein, silane and contaminants. For the latter, extreme poles of hydrocarbons and polysaccharides were taken as models.

Silanization increased markedly the retention of glucose oxidase, whether or not using a linker expected to couple the enzyme with the silane. Direct retention of the enzyme by the silane may be attributed to electrostatic attractions with the protonated amine groups (Jasienak et al., 2009). It is thus demonstrated, with the same biomolecule, that its retention by using an APTES-treated substrate and a linker does not infer covalent grafting through the linker. The occurrence of covalent binding might possibly be evaluated by examining the retention upon aging in electrolytes, keeping in mind that the silane layer itself may alter in these conditions (Dekeyser et al., 2008).

The thickness of the organic adlayer was of the order of 3 nm. In all cases, the concentration of contaminants exceeded the concentration of silane and protein. Angle-resolved measurements did not reveal any stratification in the organic adlayer, but this was not conclusive since the effect of a stratification may be masked owing to the roughness created by nanoparticles of inorganic oxyhydroxide present at the stainless steel surface. The presence of organic contaminants is unavoidable when high energy materials are exposed to air for a few hours or to vacuum for a very short time. However the influence of contaminants on the desired surface reactions is not known and is difficult to establish. The best to reduce contamination is to thoroughly clean the substrate by oxidation, to minimize the time of contact of substrates with air at any stage, and to strictly avoid outgassing. Water contact angle measurement is the best way to assess the cleanliness of the native substrate but the information will not be unambiguous after silanization and further treatments.

6. Acknowledgments

The authors thank Simon Degand for his help in statistics. They acknowledge the support of the "Conseil Régional de Picardie" (France) and the National Fondation for Scientific Research (F.N.R.S. – Belgium).

Silanization with APTES for Controlling the Interactions Between Stainless Steel and
Biocomponents: Reality vs Expectation

135

7. References

Ahimou, F., Boonaert, C.J.P., Adriaensen, Y., Jacques, P., Thonart, P., Paquot, M. & Rouxhet, P.G. (2007) XPS analysis of chemical functions at the surface of Bacillus subtilis, *Journal of Colloid and Interface Science*,Vol. 309, No. 1, pp. 49-55.

Asenath Smith, E. & Chen, W. (2008) How To Prevent the Loss of Surface Functionality Derived from Aminosilanes, *Langmuir*,Vol. 24, No. 21, pp. 12405-12409.

Balasundaram, G., Sato, M. & Webster, T.J. (2006) Using hydroxyapatite nanoparticles and decreased crystallinity to promote osteoblast adhesion similar to functionalizing with RGD, *Biomaterials*,Vol. 27, No. 14, pp. 2798-2805.

Bartouilh de Taillac, L., Porté-Durrieu, M.C., Labrugère, C., Bareille, R., Amédée, J. & Baquey, C. (2004) Grafting of RGD peptides to cellulose to enhance human osteoprogenitor cells adhesion and proliferation, *Composites Science and Technology*, Vol. 64, No. 6, pp. 827-837.

Boucher, C., Liberelle, B.t., Jolicoeur, M., Durocher, Y. & De Crescenzo, G. (2009) Epidermal growth factor tethered through coiled-coil interactions induces cell surface receptor phosphorylation, *Bioconjugate Chemistry*, Vol. 20, No. 8, pp. 1569-1577.

Briggs, D. & Seah, M.P. (1990) *Practical Surface Analysis, Auger and X-ray Photoelectron Spectroscopy* (Chichester, UK, Wiley).

Caillou, S., Gerin, P.A., Nonckreman, C.J., Fleith, S., Dupont-Gillain, C.C., Landoulsi, J., Pancera, S.M., Genet, M.J. & Rouxhet, P.G. (2008) Enzymes at solid surfaces: Nature of the interfaces and physico-chemical processes, *Electrochimica Acta*, Vol. 54, No. 1, pp. 116-122.

Charbonneau, C., Liberelle, B., Hébert, M.-J., De Crescenzo, G. & Lerouge, S. (2011) Stimulation of cell growth and resistance to apoptosis in vascular smooth muscle cells on a chondroitin sulfate/epidermal growth factor coating, *Biomaterials*, Vol. 32, No. 6, pp. 1591-1600.

Cho, Y. & Ivanisevic, A. (2004) SiOx Surfaces with lithographic features composed of a TAT peptide, *The Journal of Physical Chemistry B*, Vol. 108, No. 39, pp. 15223-15228.

Chuang, Y.-J., Huang, J.-W., Makamba, H., Tsai, M.-L., Li, C.-W. & Chen, S.-H. (2006) Electrophoretic mobility shift assay on poly(ethylene glycol)-modified glass microchips for the study of estrogen responsive element binding, *Electrophoresis*, Vol. 27, No. 21, pp. 4158-4165.

Davis, D.H., Giannoulis, C.S., Johnson, R.W. & Desai, T.A. (2002) Immobilization of RGD to <1 1 1> silicon surfaces for enhanced cell adhesion and proliferation, *Biomaterials*, Vol. 23, No. 19, pp. 4019-4027.

Dekeyser, C.M., Buron, C.C., Mc Evoy, K., Dupont-Gillain, C.C., Marchand-Brynaert, J., Jonas, A.M. & Rouxhet, P.G. (2008) Oligo(ethylene glycol) monolayers by silanization of silicon wafers: Real nature and stability, *Journal of Colloid and Interface Science*, Vol. 324, No. 1-2, pp. 118-126.

Doh, J. & Irvine, D.J. (2006) Immunological synapse arrays: Patterned protein surfaces that modulate immunological synapse structure formation in T cells, *Proceedings of the National Academy of Sciences*, Vol. 103, No. 15, pp. 5700-5705.

Dufrêne, Y.F. & Rouxhet, P.G. (1996) Surface composition, surface properties, and adhesiveness of Azospirillum brasilense - variation during growth, *Canadian Journal of Microbiology*, Vol. 42, pp. 548-556.

Dupont, I., Féron, D. & Novel, G. (1998) Effect of glucose oxidase activity on corrosion potential of stainless steels in seawater, *International Biodeterioration & Biodegradation*, Vol. 41, No. 1, pp. 13-18.

El-Ghannam, A.R., Ducheyne, P., Risbud, M., Adams, C.S., Shapiro, I.M., Castner, D., Golledge, S. & Composto, R.J. (2004) Model surfaces engineered with nanoscale roughness and RGD tripeptides promote osteoblast activity, *Journal of Biomedical Materials Research Part A*, Vol. 68A, No. 4, pp. 615-627.

Genet, M.J., Dupont-Gillain, C.C. & Rouxhet, P.G. (2008) XPS analysis of biosystems and biomaterials, in: M. E. (Ed) *Medical Applications of Colloids* (New York, Springer), pp. 177-307.

Gooding, J.J. & Ciampi, S. (2011) The molecular level modification of surfaces: from self-assembled monolayers to complex molecular assemblies, *Chemical Society Reviews*, DOI: 10.1039/c0cs00139b.

Haensch, C., Hoeppener, S. & Schubert, U.S. (2010) Chemical modification of self-assembled silane based monolayers by surface reactions, *Chemical Society Reviews*, Vol. 39, No. 6, pp. 2323-2334.

Hanawa T. 2002. Metallic biomaterials. In: Ikada Y, editor. Recent research and developments in biomaterials, Research Signpost, p. 11-31.

Howarter, J.A. & Youngblood, J.P. (2006) Optimization of silica silanization by 3-Aminopropyltriethoxysilane, *Langmuir*, Vol. 22, No. 26, pp. 11142-11147.

Iucci, G., Dettin, M., Battocchio, C., Gambaretto, R., Bello, C.D. & Polzonetti, G. (2007) Novel immobilizations of an adhesion peptide on the TiO2 surface: An XPS investigation, *Materials Science and Engineering: C*, Vol. 27, No. 5-8, pp. 1201-1206.

Jasienak, M., Suzuki, S., Montero, M., Wentrup-Byrne, E., Griesser, H.J. & Grondahl, L. (2009) Time-of-flight secondary ion mass spectrometry study of the orientation of a bifunctional diblock copolymer attached to a solid substrate, *Langmuir*, Vol. 25, No. 2, pp. 1011-1019.

Jin, L., Horgan, A. & Levicky, R. (2003) Preparation of end-tethered DNA monolayers on siliceous surfaces using heterobifunctional cross-linkers, *Langmuir*, Vol. 19, No. 17, pp. 6968-6975.

Katsikogianni, M.G. & Missirlis, Y.F. (2010) Interactions of bacteria with specific biomaterial surface chemistries under flow conditions, *Acta Biomaterialia*, Vol. 6, No. 3, pp. 1107-1118.

Killian, M.S., Wagener, V., Schmuki, P. & Virtanen, S. (2010) Functionalization of metallic magnesium with protein layers via linker molecules, *Langmuir*, Vol. 26, No. 14, pp. 12044-12048.

Kim, J., Cho, J., Seidler, P.M., Kurland, N.E. & Yadavalli, V.K. (2010) Investigations of chemical modifications of amino-terminated organic films on silicon substrates and controlled protein immobilization, *Langmuir*, Vol. 26, No. 4, pp. 2599-2608.

Kim, J., Seidler, P., Wan, L.S. & Fill, C. (2009a) Formation, structure, and reactivity of amino-terminated organic films on silicon substrates, *Journal of Colloid and Interface Science*, Vol. 329, No. 1, pp. 114-119.

Kim, W.-J., Kim, S., Lee, B.S., Kim, A., Ah, C.S., Huh, C., Sung, G.Y. & Yun, W.S. (2009b) Enhanced protein immobilization efficiency on a TiO2 surface modified with a hydroxyl functional group, *Langmuir*, Vol. 25, No. 19, pp. 11692-11697.

Silanization with APTES for Controlling the Interactions Between Stainless Steel and
Biocomponents: Reality vs Expectation

137

Koh, I., Wang, X., Varughese, B., Isaacs, L., Ehrman, S.H. & English, D.S. (2006) Magnetic iron oxide nanoparticles for biorecognition: Evaluation of surface coverage and activity, *The Journal of Physical Chemistry B*, Vol. 110, No. 4, pp. 1553-1558.

Kohli, P., Taylor, K.K., Harris, J.J. & Blanchard, G.J. (1998) Assembly of covalently-coupled disulfide multilayers on gold, *Journal of the American Chemical Society*, Vol. 120, No. 46, pp. 11962-11968.

Landoulsi, J., Dagbert, C., Richard, C., Sabot, R., Jeannin, M., El Kirat, K. & Pulvin, S. (2009) Enzyme-induced ennoblement of AISI 316L stainless steel: Focus on pitting corrosion behavior, *Electrochimica Acta*, Vol. 54, No. 28, pp. 7401-7406.

Landoulsi, J., Genet, M.J., Richard, C., El Kirat, K., Pulvin, S. & Rouxhet, P.G. (2008a) Evolution of the passive film and organic constituents at the surface of stainless steel immersed in fresh water, *Journal of Colloid and Interface Science*, Vol. 318, No. 2, pp. 278-289.

Landoulsi, J., Genet, M.J., Richard, C., El Kirat, K., Rouxhet, P.G. & Pulvin, S. (2008b) Ennoblement of stainless steel in the presence of glucose oxidase: Nature and role of interfacial processes, *Journal of Colloid and Interface Science*, Vol. 320, No. 2, pp. 508-519.

Landoulsi, J., Kirat, K.E., Richard, C., Féron, D. & Pulvin, S. (2008c) Enzymatic approach in microbial-influenced corrosion: A review based on Stainless Steels in Natural Waters, *Environmental Science & Technology*, Vol. 42, No. 7, pp. 2233-2242.

Lapin, N.A. & Chabal, Y.J. (2009) Infrared characterization of biotinylated silicon oxide surfaces, surface stability, and specific attachment of streptavidin, *The Journal of Physical Chemistry B*, Vol. 113, No. 25, pp. 8776-8783.

Le Bozec, N., Compère, C., L'Her, M., Laouenan, A., Costa, D. & Marcus, P. (2001) Influence of stainless steel surface treatment on the oxygen reduction reaction in seawater, *Corrosion Science*, Vol. 43, No. 4, pp. 765-786.

Li, G., Yang, P., Qin, W., Maitz, M.F., Zhou, S. & Huang, N. (2011) The effect of coimmobilizing heparin and fibronectin on titanium on hemocompatibility and endothelialization, *Biomaterials*, Vol. 32, No. 21, pp. 4691-4703.

Libertino, S., Giannazzo, F., Aiello, V., Scandurra, A., Sinatra, F., Renis, M. & Fichera, M. (2008) XPS and AFM characterization of the enzyme glucose oxidase immobilized on SiO$_2$ surfaces, *Langmuir*, Vol. 24, No. 5, pp. 1965-1972.

Ma, L., Zhou, J., Gao, C. & Shen, J. (2007) Incorporation of basic fibroblast growth factor by a layer-by-layer assembly technique to produce bioactive substrates, *Journal of Biomedical Materials Research Part B: Applied Biomaterials*, Vol. 83B, No. 1, pp. 285-292.

Mantel, M., Rabinovich, Y.I., Wightman, J.P. & Yoon, R.H. (1995) A Study of Hydrophobic Interactions between Stainless Steel and Silanated Glass Surface Using Atomic Force Microscopy, *Journal of Colloid and Interface Science*, Vol. 170, No. 1, pp. 203-214.

Martin, H.J., Schulz, K.H., Bumgardner, J.D. & Walters, K.B. (2007) XPS Study on the Use of 3-Aminopropyltriethoxysilane to Bond Chitosan to a Titanium Surface, *Langmuir*, Vol. 23, No. 12, pp. 6645-6651.

Matinlinna, J., P., Lassila, L., V. J., Özcan, M., Yli-Urpo, A. & Vallittu, P., K. (2004) An introduction to silanes and their clinical applications in dentistry, *International Journal of Prosthodontics*, Vol. 17, No. 211, pp. 155-164.

Matinlinna, J.P. & Vallittu, P.K. (2007) Silane based concepts on bonding resin composite to metals, *The Journal of Contemporary Dental Practice*, Vol. 8, No. 2, pp. 1-8.

Meng, S., Liu, Z., Shen, L., Guo, Z., Chou, L.L., Zhong, W., Du, Q. & Ge, J. (2009) The effect of a layer-by-layer chitosan-heparin coating on the endothelialization and coagulation properties of a coronary stent system, *Biomaterials*, Vol. 30, No. 12, pp. 2276-2283.

Minier, M., Salmain, M.l., Yacoubi, N., Barbes, L., Méthivier, C., Zanna, S. & Pradier, C.-M. (2005) Covalent Immobilization of Lysozyme on Stainless Steel. Interface spectroscopic characterization and measurement of enzymatic activity, *Langmuir*,Vol. 21, No. 13, pp. 5957-5965.

Mosse, W.K.J., Koppens, M.L., Gengenbach, T.R., Scanlon, D.B., Gras, S.L. & Ducker, W.A. (2009) Peptides grafted from solids for the control of interfacial properties, *Langmuir*, Vol. 25, No. 3, pp. 1488-1494.

Müller, R., Abke, J., Schnell, E., Macionczyk, F., Gbureck, U., Mehrl, R., Ruszczak, Z., Kujat, R., Englert, C., Nerlich, M. & Angele, P. (2005) Surface engineering of stainless steel materials by covalent collagen immobilization to improve implant biocompatibility, *Biomaterials*, Vol. 26, No. 34, pp. 6962-6972.

Nanci, A., Wuest, J.D., Peru, L., Brunet, P., Sharma, V., Zalzal, S. & McKee, M.D. (1998) Chemical modification of titanium surfaces for covalent attachment of biological molecules, *Journal of Biomedical Materials Research*, Vol. 40, No. 2, pp. 324-335.

NIST X-ray Photoelectron Spectroscopy Database, NIST Standard Reference Database 20, Version 3.4 (Web Version), http://srdata.nist.gov/xps/index.htm.

North, S.H., Lock, E.H., Cooper, C.J., Franek, J.B., Taitt, C.R. & Walton, S.G. (2004) Plasma-based surface modification of polystyrene microtiter plates for covalent immobilization of biomolecules, *ACS Applied Materials & Interfaces*, Vol. 2, No. 10, pp. 2884-2891.

Olefjord, I. & Wegrelius, L. (1996) The influence of nitrogen on the passivation of stainless steels, *Corrosion Science*, Vol. 38, No. 7, pp. 1203-1220.

Palestino, G., Agarwal, V., Aulombard, R., Peí rez, E.a. & Gergely, C. (2008) Biosensing and protein fluorescence enhancement by functionalized porous silicon devices, *Langmuir*, Vol. 24, No. 23, pp. 13765-13771.

Pasternack, R.M., Rivillon Amy, S. & Chabal, Y.J. (2008) Attachment of 3-(Aminopropyl)triethoxysilane on silicon oxide surfaces: Dependence on solution temperature, *Langmuir*, Vol. 24, No. 22, pp. 12963-12971.

Plueddemann, E.W. (1991) *Silane Coupling Agents* (New York, Plenum).

Porté-Durrieu, M.C., Guillemot, F., Pallu, S., Labrugère, C., Brouillaud, B., Bareille, R., Amédée, J., Barthe, N., Dard, M. & Baquey, C. (2004) Cyclo-(DfKRG) peptide grafting onto Ti-6Al-4V: physical characterization and interest towards human osteoprogenitor cells adhesion, *Biomaterials*, Vol. 25, No. 19, pp. 4837-4846.

Puleo, D.A. (1997) Retention of enzymatic activity immobilized on silanized Co-Cr-Mo and Ti-6Al-4V, *Journal of Biomedical Materials Research*, Vol. 37, pp. 222-228.

Quan, D., Kim, Y. & Shin, W. (2004) Characterization of an amperometric laccase electrode covalently immobilized on platinum surface, *Journal of Electroanalytical Chemistry*, Vol. 561, pp. 181-189.

Ratner B.D., Hoffman A.S., Schoen F.J. & Lemons J.E. (Eds) (2004) *Biomaterials Science: An Introduction to Materials in Medecine* (Academic Press, San Diego, U.S.A., 2d Ed.).

Silanization with APTES for Controlling the Interactions Between Stainless Steel and
Biocomponents: Reality vs Expectation

139

Rouxhet, P.G. & Genet, M.J. (2011) XPS analysis of bio-organic systems, *Surface and Interface Analysis*. (In press).

Rouxhet, P.G., Misselyn-Bauduin, A.M., Ahimou, F., Genet, M.J., Adriaensen, Y., Desille, T., Bodson, P. & Deroanne, C. (2008) XPS analysis of food products: toward chemical functions and molecular compounds, *Surface and Interface Analysis*,Vol. 40, No. 3-4, pp. 718-724.

Sarath Babu, V.R., Kumar, M.A., Karanth, N.G. & Thakur, M.S. (2004) Stabilization of immobilized glucose oxidase against thermal inactivation by silanization for biosensor applications, *Biosensors and Bioelectronics*,Vol. 19, No. 10, pp. 1337-1341.

Sargeant, T.D., Rao, M.S., Koh, C.-Y. & Stupp, S.I. (2008) Covalent functionalization of NiTi surfaces with bioactive peptide amphiphile nanofibers, *Biomaterials*,Vol. 29, No. 8, pp. 1085-1098.

Sasou, M., Sugiyama, S., Yoshino, T. & Ohtani, T. (2003) Molecular flat mica surface silanized with methyltrimethoxysilane for fixing and straightening DNA, *Langmuir*, Vol. 19, No. 23, pp. 9845-9849.

Schuessele, A., Mayr, H., Tessmar, J. & Goepferich, A. (2009) Enhanced bone morphogenetic protein-2 performance on hydroxyapatite ceramic surfaces, *Journal of Biomedical Materials Research Part A*, Vol. 90A, No. 4, pp. 959-971.

Scofield, J.H. (1976) Hartree-Slater subshell photoionization cross-sections at 1254 and 1487 eV, *Journal of Electron Spectroscopy and Related Phenomena*, Vol. 8, No. 2, pp. 129-137.

Siperko, L.M., Jacquet, R. & Landis, W.J. (2006) Modified aminosilane substrates to evaluate osteoblast attachment, growth, and gene expression in vitro, *Journal of Biomedical Materials Research Part A*, Vol. 78A, No. 4, pp. 808-822.

Son, K.J., Ahn, S.H., Kim, J.H. & Koh, W.-G. (2011) Graft copolymer-templated mesoporous TiO_2 films micropatterned with Poly(ethylene glycol) hydrogel: Novel platform for highly sensitive protein microarrays, *ACS Applied Materials & Interfaces*,Vol. 3, No. 2, pp. 573-581.

Sordel, T., Kermarec-Marcel, F., Garnier-Raveaud, S., Glade, N., Sauter-Starace, F., Pudda, C., Borella, M., Plissonnier, M., Chatelain, F., Bruckert, F. & Picollet-D'hahan, N. (2007) Influence of glass and polymer coatings on CHO cell morphology and adhesion, *Biomaterials*,Vol. 28, No. 8, pp. 1572-1584.

Subramanian, A., Kennel, S.J., Oden, P.I., Jacobson, K.B., Woodward, J. & Doktycz, M.J. (1999) Comparison of techniques for enzyme immobilization on silicon supports, *Enzyme and Microbial Technology*,Vol. 24, No. 1-2, pp. 26-34.

Suzuki, N. & Ishida, H. (1996) A review on the structure and characterization techniques of silane/matrix interphases, *Macromolecular Symposia*,Vol. 108, No. 1, pp. 19-53.

Suzuki, S., Whittaker, M.R., Grøndahl, L., Monteiro, M.J. & Wentrup-Byrne, E. (2006) Synthesis of soluble phosphate polymers by RAFT and their in vitro mineralization, *Biomacromolecules*, Vol. 7, No. 11, pp. 3178-3187.

Tanuma, S., Powell, C.J. & Penn, D.R. (1997) Calculations of electron inelastic mean free paths (IMFPs) VI. Analysis of the gries inelastic scattering model and predictive IMFP equation, *Surface and Interface Analysis*, Vol. 25, No. 1, pp. 25-35.

Tesson, B., Genet, M.J., Fernandez, V., Degand, S., Rouxhet, P.G. & Martin-Jézéquel, V. (2009) Surface chemical composition of diatoms, *ChemBioChem*, Vol. 10, No. 12, pp. 2011-2024.

Toworfe, G.K., Bhattacharyya, S., Composto, R.J., Adams, C.S., Shapiro, I.M. & Ducheyne, P. (2009) Effect of functional end groups of silane self-assembled monolayer surfaces on apatite formation, fibronectin adsorption and osteoblast cell function, *Journal of Tissue Engineering and Regenerative Medicine*, Vol. 3, No. 1, pp. 26-36.

Toworfe, G.K., Composto, R.J., Shapiro, I.M. & Ducheyne, P. (2006) Nucleation and growth of calcium phosphate on amine-, carboxyl- and hydroxyl-silane self-assembled monolayers, *Biomaterials*, Vol. 27, No. 4, pp. 631-642.

Vaidya, A.A. & Norton, M.L. (2004) DNA attachment chemistry at the flexible silicone elastomer surface: Toward disposable microarrays, *Langmuir*, Vol. 20, No. 25, pp. 11100-11107.

Weetall, H.H. (1993) Preparation of immobilized proteins covalently coupled through silane coupling agents to inorganic supports. *Applied Biochemistry and Biotechnology*, Vol. 41, No. 157, pp. 157-188.

Weng, Y.J., Hou, R.X., Li, G.C., Wang, J., Huang, N. & Liu, H.Q. (2008) Immobilization of bovine serum albumin on TiO_2 film via chemisorption of H_3PO_4 interface and effects on platelets adhesion, *Applied Surface Science*, Vol. 254, No. 9, pp. 2712-2719.

Williams, R. (2010) *Surface Modifications of Biomaterials: Methods Analysis and Applications* (Woodhead Publishing Ltd).

Wink, T., van Zuilen, S.J., Bult, A. & van Bennekom, W.P. (1997) Self-assembled monolayers for biosensors. *Analyst*, Vol 122, pp. 43R-50R

Xiao, S.J., Textor, M., Spencer, N.D., Wieland, M., Keller, B. & Sigrist, H. (1997) Immobilization of the cell-adhesive peptide Arg–Gly–Asp–Cys (RGDC) on titanium surfaces by covalent chemical attachment, *Journal of Materials Science: Materials in Medicine*, Vol. 8, No. 12, pp. 867-872.

Xie, Y., Hill, C.A.S., Xiao, Z., Militz, H. & Mai, C. (2010) Silane coupling agents used for natural fiber/polymer composites: A review, *Composites Part A: Applied Science and Manufacturing*, Vol. 41, No. 7, pp. 806-819.

Yoshioka, T., Tsuru, K., Hayakawa, S. & Osaka, A. (2003) Preparation of alginic acid layers on stainless-steel substrates for biomedical applications, *Biomaterials*, Vol. 24, No. 17, pp. 2889-2894.

Zhou, G.-T., Yao, Q.-Z., Wang, X. & Yu, J.C. (2006) Preparation and characterization of nanoplatelets of nickel hydroxide and nickel oxide, *Materials Chemistry and Physics*, Vol. 98, No. 2-3, pp. 267-272.

Zile, M.A., Puckett, S. & Webster, T.J. Nanostructured titanium promotes keratinocyte density, *Journal of Biomedical Materials Research Part A*, Vol. 97A, No. 1, pp. 59-65.

7

Comparative Metal Ion Binding to Native and Chemically Modified *Datura innoxia* Immobilized Biomaterials

Gary D. Rayson and Patrick A. Williams
Department of Chemistry and Biochemistry
New Mexico State University, Las Cruces, NM
USA

1. Introduction

Removal of toxic heavy metal ions from contaminated water is required to provide safe drinking water. This can be effected either within the waste stream at contaminate source or at point of use. Incorporation of remediation technologies at either location requires the removal of pollutants at parts per million and parts per billion concentrations from water containing more benign metal ions (e.g., Ca^{2+}, Mg^{2+}, and Na^+) at concentrations three to six orders of magnitude higher. Materials derived from plants or microorganisms (e.g., algae and fungi) have been shown to enable the reduction of trace concentrations of heavy metal ions to below regulatory limits (Davis, et al., 2003).

Such nonliving biomaterials have been reported to have exhibit high capacity, rapid binding, and selectivity towards heavy metals (Drake and Rayson, 1996). It is postulated that functional groups native to the lipids, carbohydrates, and proteins found in the cell walls of the biomaterial are responsible for uptake (biosorption) of metal ions (Gardea-Torresdey, et al., 1999; 2001; Drake and Rayson, 1996; Drake, et al., 1997; Kelley, et al., 1999). For biomaterials to become a commercially viable method of metal remediation and recovery these functional groups must be identified and their contribution to overall metal binding capacity quantified. Knowledge of such informaiton would allow either simple chemical alteration to the biomaterial, allowing for targeting of specific metals, or an enhancement of biomaterial metal binding.

Significant progress has been made to identify the chemical functionalities involved in the biosorption of numerous metal ions by a variety of plant and algal tissues (Gardea-Torresdey, et al., 1999; 2001; Riddle, et al., 2002; Fourest and Volesky, 1996; Drake, et al., 1997;Jackson, et al., 1993). Several techniques have been reported to probe local chemical environments of biosorbed metal ions. These have included X-ray absorption (Gardea-Torresdey, et al., 1999; Riddle, et al., 2002), lanthanide luminescence (Drake, et al., 1997; Serna, et al., 2010), and metal NMR (Xia and Rayson, 1996; 2002; Kelley, et al., 1999; Majidi, et al., 1990) spectroscopy. Analysis of total metal ion binding isotherm data modeling (Volesky, 2000) has also been described. Efforts to address the chemical heterogeneity of those biosorbed materials have also employed regularized regression analysis of isotherm data (Lin, et al., 1996). Additionally, these chemical intensities have been studied through selective removal of binding moieties by their reactive modification (Drake, et al., 1996).

It has been shown that carboxyl groups present in the cell walls of nonliving biomaterials contribute to metal-ion binding (Gardea-Torresdey, et al., 1999; 2001; Riddle, et al., 1997; Kelley, et al., 1999). Our group has used a variety of methods(Lin, et al., 2002; Drake, et al., 1996; Xia and Rayson, 1995; Drake, et al., 1996; 1997) to characterize the binding groups present in the cell walls of *Datura innoxia*. This plant is a member of the *Solanaceae* plant family and native to Mexico and the southwestern United States. To minimize variability of cell types investigated, the cell-wall fragments used are cultured anther cells of the plant. This plant was selected for study because it is a heavy metal resistant perennial that is both tolerant of arid climates and resistant to herbivory (Drake et al., 1996).

Our group has concentrated primarily on nonviable biomaterials, specifically cell wall fragments from the cultured anther cells of *Datura innoxia*. The present study used frontal affinity chromatography with inductively coupled plasma optical emission spectroscopy (ICP-OES) detection for simultaneous monitoring both uptake and release of metal ions to both a chemically modified and native *D. innoxia* biomaterial (Williams and Rayson, 2003). The objective of the present study was to further investigate such sites through sequential exposure and subsequent stripping of three similar metal ions (Cd^{2+}, Ni^{2+}, and Zn^{2+}) to both a modified and the native biosorbents, thus to study the role of carboxylate furface functionalities on passive metal ion binding of this material.

It has been demonstrated (Drake, et al., 1996) that carboxylate-containing binding sites can be removed through the formation of the corresponding methyl esters by reaction with acidic methanol for 72 hours (Drake, et al., 1996). Undertaking a similar series of experiments with such a chemically modified sorbent enables the investigation of alternate binding sites.

2. Materials and methods

2.1 Esterification of biomaterial
The cultured anther cells from *D. innoxia* were washed and prepared as described elsewhere (Drake, et al,, 1996; 1997). Only cell fragment aggregates with a mesh size greater than 200 (< 127 μm) were used for esterification. Following a method described elsewhere (Drake, et al., 1996), 10.0 grams of the biomaterial were suspended in 0.1 M HCl in methanol. The slurry was continuously heated at 60°C and stirred for 72 hours. The biomaterial was then recovered through vacuum filtration, rinsed three times with 16.0-MΩ water (Barnstead,Millipore Ultrapure), freeze-dried, and set aside for later immobilization.

2.2 Immobilization of biomaterial
In their native state, biomaterials have poor mechanical strength, low density, and a small particle size that can cause column clogging (Stark and Rayson, 2000). These characteristics can yield poor candidates for column-based water treatment applications. For this study native and modified *D. innoxia* biomaterials were each immobilized in a polysilicate matrix. The 40-60-mesh size fraction of the ground, and sieved immobilized biosorbents was then packed into columns. The process for immobilization has been described in detail elsewhere (Stark and Rayson, 2000).

Briefly, a suspension of 20 grams of the 100-200 mesh fraction of the washed biomaterial was generated with 300 mL of 5% v/v sulfuric acid adjusted to pH 2.0 by addition of a 6% (w/v) solution of $Na_2SiO_3 \bullet 5H_2O$. This suspension was stirred for 1 hour and the pH of the solution further raised to 7.0 by incremental addition of the Na_2SiO_3 solution. A gelatinous polymer formed at pH 7.0. The solution was stirred an additional 30 minutes, covered and stored at 4°C overnight. The resulting aqueous layer was removed. Excess sulfate ions were removed from the gel by successive washings with distilled deionized water until the aqueous phase failed a sulfate test using a few drops of a 1.0% barium solution (as the nitrate salt). When no precipitate was visible, one final wash with the distilled deionized water was performed to ensure complete sulfate removal.

Remaining polymer was transferred to ceramic evaporating dishes and baked at approximately 100° C until completely dry. The immobilized *D. innoxia* biomaterial was then ground and sieved. The 40-60-mesh (423-635 μm) particle size fraction was collected. Percent compositions of cell wall biomaterial were determined gravimetrically to be 64.6% and 75.3% for the native and modified materials, respectively.

2.3 Biomaterial columns

The columns used have been described elsewhere (Williams and Rayson, 2003) and were constructed in-house from Plexiglas™ tubing (2.5 cm in length and 3 mm i.d.). Teflon™ tubing (0.8-mm i.d.) was used for all column connections. Interface of the column to the ICP-OES was accomplished by connecting the column outlet directly to the inlet of the cross-flow type nebulizer using the minimum length of Teflon™ tubing (15 cm). Column effluent was monitored for each of 27 different metals simultaneously. Table 1 list the elements observed and their respective emission wavelengths.

Element	Wavelength/nm	Element	Wavelength/nm	Element	Wavelength/nm
Na	589.00	Mg	279.00	Al	396.15
Ca	317.90	Cr	267.70	Mn	257.60
Fe	259.90	Ni	231.60	Cu	324.70
Zn	213.80	Cd	228.80	Ag	328.00
Sn	189.90	Pb	220.30	Ba	493.40
Sr	421.50	U	409.00	Y	371.00
V	242.40	Mo	202.00	Co	228.60
Si	251.60	As	193.60	Se	196.00
Tb	350.97	Eu	381.97	Th	283.73

Table 1. Elements and the corresponding emission wavelength used during monitoring of column effluents (elements of interests in this study indicated by boldface print).

Each column was packed with approximately 125 mg of the immobilized *D. innoxia* material and flow tested using distilled deionized water. Once packed and tested for leaks, each column was exposed to 20 mL of 1.0-M HCl using a peristaltic pump (Model Rabbit, Rainin) (1.0 mL/min for 20 min) and the effluent monitored for metals released from the biomaterial. Following the acid rinse, the columns were then exposed to 5 mL of distilled deionized water (1.0 mL/min for five minutes) to reestablish an ambient pH influent environment (~pH 6.2).

These studies involved, initially, the exposure of a small column (3.0 mm i.d., 10.0 mm in length) to an equimolar mixture of metal ions, specifically, Cd^{2+}, Zn^{2+}, and Ni^{2+}, and exposure to solutions of each metal sequentially while continuously monitoring these (and other) metal species in the column effluent.

2.4 Frontal affinity chromatography with inductively coupled plasma atomic emission detection

This technique has been described in detail elsewhere (Williams and Rayson, 2003). Briefly, the biomass packed column, having been exposed to 20 mL of 1.0M HCl to remove any metals remaining on the biomaterial (effluent monitored by ICPAES), was exposed to 5 mL of 16 MΩ, distilled-deionized water. The influent was a metal-ion solution, 0.1mM-0.2mM, made from the nitrate salt of Cd^{2+}, Ni^{2+}, or Zn^{2+}. Initially, the influent metal ion concentration increased as a step function.

Each influent was pumped through a column using a peristaltic pump (Rainin) at the rate of ~1.0 mL min^{-1} to a cross-flow type nebulizer and Scott-type double-pass spray chamber of the ICP-OES spectrometer (Jarrell-Ash, AtomComp700). The biomaterial in each column was exposed to each metal solution for 50 minutes. The effluent was monitored and resulting break-through curves were recorded for each metal ion (Figures 1A-C). Following exposure to the column, bound metal ions were stripped from the column using each of two exposures to a 1.0 M HCl solution. The first 150-second (~2.5 mL) exposure removed approximately 98% of the metal ions on the column (Figure 1D). The second 20 minute (~20 mL) exposure removed the remaining 2%. This was followed by a 5 minute (~5 mL) rinse with distilled deionized water to return the pH to f the biomaterial to that of the natural water (~6.2). Influent pH was not buffered to a predetermined pH to more accurately emulate conditions of a natural water supply within a remediation application.

With a three metal system there are six combinations that the metals can be sequentially exposed to the biomaterial (CdZnNi, CdNiZn, NiCdZn, NiZnCd, ZnCdNi, and ZnNiCd) and all six were performed on each column. Specifically, the ZnNiCd sequence involved exposure of a column packed with a biosorbents to a 0.20 mM Zn^{2+}solution for 50 minutes (Figure 1A). The influent was changed to a 0.20 mM Ni^{2+} solution for another 50 minutes (Figure 1B). Similarly, a 0.20 mM Cd^{2+} solution was pumped through the same column for an additional 50 minutes (Figure 1C). The column was then exposed to 1.0 M HCl for 2.5 and 20 minutes to remove all bound metal ions (Figure 1D).

Simultaneous exposure of the three metals at the same molar concentration was also undertaken for each column with both the native (Figure 2) and modified (not shown) biomaterials. All determinations were performed in triplicate with three separate columns packed with each individual biosorbent.

3. Results and discussion

3.1 Binding capacities of native and modified *d. innoxia*

Figure 1 illustrates a sequential 50-mL exposure of 0.1mM Zn^{2+}, Ni^{2+}, and Cd^{2+} to the native *D. innoxia*. Figure 2 shows the sequential 50-mL exposures of 0.2mM Zn^{2+}, Ni^{2+}, and Cd^{2+} to the modified biomaterial. Table 2 lists the observed binding capacities for the three metals to the modified *D. innoxia* reported in moles metal bound per gram of biomaterial for each replicate. The capacities reported are mass balance capacities, as the column effluent was monitored.

Simple statistical analysis using a t-test confirms the binding capacity of each metal to the biomaterial was decreased as expected. The position of the metal in the sequence was similarly indicated as affecting capacities in both the native and the modified biomaterial materials. Further statistical analysis indicated that not only the position in the sequence but the history of exposure appears to impact apparent steady state binding metal ion capacities of the biomaterial.

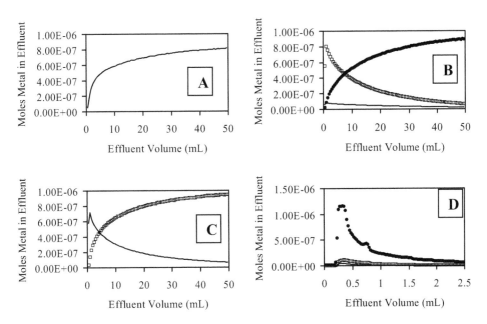

Fig. 1. Effluent profiles resulting from solutions of (A) 0.2mM Zn^{2+} (—), (B) followed by 0.2 mM Ni^{2+} (□), followed by 0.2mM Cd^{2+}(•) pumped through a column packed with chemically modified *D. innoxia*.(D) Effluent profile of first 1.0M HCl wash of the metal-laden material.

3.2 Modified material three-metal sequential study

Total amounts of metal bound to the native and modified materials at each stage of the sequence are listed in Table 2. Although the average amount of the fist metal ion bound was 24.5 µmol g^{-1}, the average total metal ion bound after exposure to the second and third metals was 31.0 µmol g^{-1}. This increase in the amount of metal captured by the biomass could be indicative of either the presence of metal ion-specific binding sites, or some degree of competitive metal ion binding.

Simultaneous exposure of the biosorbent material to an equimolar solution of all three metal-ions (0.2 mM) was undertaken. Figure 2A shows the resulting effluent concentration profile. Even with a total metal ion concentration of 0.6 mM, the effluent concentration maximized at only 94% of the influent concentration (i.e., 0.19 mM). The modified biomaterial average capacity with simultaneous exposure was 41.72 µmol g^{-1} biomaterial. Specifically, total amounts of metal ion bound were 13.24, 14.17, and 14.31 µmol g^{-1} for Ni^{2+} Zn^{2+} Cd^{2+}, respectively. This suggests no significant binding preference of the *D. innoxia* cell material for these ions.

Figure 2B shows the effluent concentration profile of the subsequent 1.0 M HCl metal-ion recovery/strip step. Total metal ion recovered from the two acid washes was 27.48 µmol g^{-1}.

This included 7.68, 10.21, and 9.59 µmol g^{-1} for Ni^{2+} Zn^{2+} Cd^{2+}, respectively. Nickel recovery was lowest among the three ions (58.0%), while cadmium and zinc recovered 67.0% and 72.1%, respectively. This suggests the affinity of some of the Ni^{2+} binding sites were higher than those for each of the other two metal ions.

Sequence	Column a	Column b	Column c	Average	Std. Dev.	%RSD
Cd	28.26	22.63	25.49	25.46	2.56	10.05
Ni	23.57	20.79	23.35	22.57	1.84	8.16
Zn	24.56	24.88	26.76	25.40	1.57	6.16
Cd Ni	27.14	27.14	24.29	26.19	1.64	6.28
Cd Zn	30.33	24.57	24.51	26.47	3.34	12.62
Ni Cd	28.39	26.04	29.95	28.13	1.97	7.01
Ni Zn	29.72	25.95	25.57	27.08	2.29	8.47
Zn Cd	27.53	24.49	31.49	27.84	3.51	12.61
Zn Ni	25.99	26.09	27.66	26.58	0.94	3.52
Cd Ni Zn	30.83	27.00	30.06	29.30	2.02	6.91
Cd Zn Ni	27.89	24.21	27.54	26.55	2.03	7.64
Ni Cd Zn	32.89	26.00	25.42	28.10	4.15	14.78
Ni Zn Cd	21.63	24.57	27.69	24.63	3.03	12.29
Zn Cd Ni	27.45	27.45	40.95	31.95	7.80	24.40
Zn Ni Cd	26.27	23.75	23.62	24.55	1.49	6.09

Table 2. Influent metal bound at each stage of the sequential exposure study. Values presented are given in µmoles metal per gram of modified biomaterial.

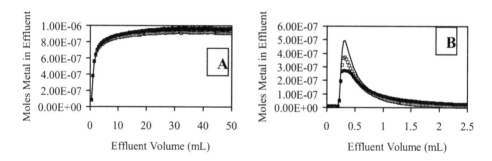

Fig. 2. Simultaneous exposure of 0.2mM Ni^{2+}(□), Cd^{2+}(•), and Zn^{2+}(—) to modified *D. innoxia* column (A) and is subsequent striping using 1.0 M HCl (B).

3.3 Comparison of native and modified biomaterials

By examining the results from the native and modified *D. innoxia* studies together, a hypothesis regarding the carboxyl sites contribution to the metal-ion binding process can be formulated. It was proposed, in previous studies of *D. innoxia*, that esterification of carboxylate sites can decrease metal uptake by as much as 40% (Drake, et al., 1996), depending on the metal.

3.3.1 Influent and total metal bound comparison

By taking the ratio of metal bound for the modified and native biomaterials at each position in the sequence and averaging those ratios for both the influent metal bound and the total metal bound, the average effect of the esterification procedure on metal binding can be quantified. On average, the influent metal-ion bound decreased by 43% while the total metal bound decreased by 54%. Figure 3 illustrates a further breakdown these comparisons by metal-ion position within the sequence. For total metal bound to the biomaterial, metal ion position had only a small effect in the percentage decrease in binding capacity observed in the modified biomaterial. With the first metal ion capacity dropping 52%, the second decreased to 54% and the third decreased to 55%. However, the influent metal ion capacity seems to be more effected by its position in the sequence. The first metal ion exposed demonstrated a decrease in its capacity of 52%, while the second and third ions exposed decreased by only 42% and 40% respectively.

Figure 4 shows additional data regarding the influent metal bound by examining both position and specific metal ion exposed in the sequence. For nickel and zinc there was a steady decrease in the observed effect of the modification as their position in the sequence moved from first to third. Nickel was most pronounced, as the effect of the modification was decreased in capacity by 54% when nickel was the first ion exposed, 41% when it is the second, and 35% when the third. Zinc was similar as it drops from a 51% decreased in binding capacity when it was the first metal exposed, a 47% decrease when it was second, to a 41% decrease when it was third. Cadmium did not follow this pattern of decreasing effect based on position. When cadmium was the first metal exposed a decrease in binding capacity of 49% is observed, a decrease in capacity of 38% was observed when it was second, and a decrease of 44% when it metal on the column. The modified biomaterial total metal bound decreased between 49 – 57% depending on influent metal and position with no apparent pattern.

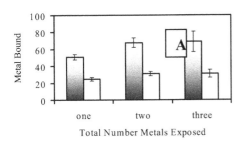 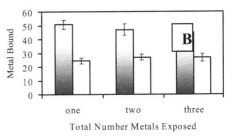

Fig. 3. Average total (A) and influent (B) metal-ion bound for native (shaded) and modified biomaterial based on number of metals exposed in series. Metal bound given in percentage metal bound.

Simultaneous exposure of the three metal-ions showed an overall decrease of 43.7% for total metal bound for the modified biomaterial. Zinc showed the largest decrease showing a 50.2% decrease in binding capacity. The modified biomaterial exhibited a loss of 43.5% for cadmium. Nickel showed the smallest effect from the modification losing 34.8% of its capacity. It should be noted that zinc, which lost the most, had the largest capacity (28.5 μmol g^{-1}), while nickel, which demonstrated the smallest effect of modification, had the smallest capacity on the native biomaterial (20.3 μmol g^{-1}).

3.3.2 Statistical analysis

Because of the numerous variables studied pertaining to the ability of this adsorbent to remove each of these metal ions from a flowing influent, it became imperative that statistical tools be employed to ascertain differences (and similarities) in metal binding. Variables that were tested include the esterification of carboxylate functionalities, the identity of the metal bond, the identity of the metal ion(s) displaced or removed, and the general history of a column of the biosorbent. The Student-t test was employed to test the hypothesis that the measured means of binding capacities between any two conditions were statistically the same. The criteria used for these test were a 95% confidence level with 2 - 5 degrees of freedom.

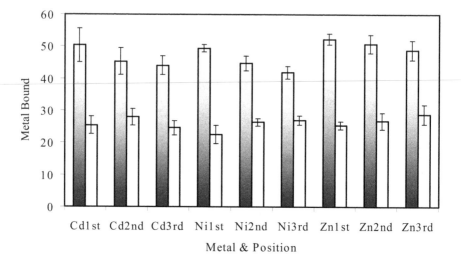

Fig. 4. Comparison of native (shaded) and modified *D. innoxia* columns influent-metal bound capacities. Metal bound is reported in micromoles per gram biomaterial.

To statistically treat the results presented above in the comparison of the native and esterified biomaterial, two methods of statistical analysis were used. The Student-t test was used to test the statistical difference between the amounts of metal bound to the biomaterial at each stage in the studies at each stage for the native and modified biomaterials. This would also reinforce the position that by undertaking a simple chemical modification, the binding properties of a biomaterial could be altered significantly. Also, the Student-t test was used to examine the impact of exposure order and history (e.g., whether the amount of cadmium bound to the biomaterial is statistically different for the series $Ni^{2+} \Rightarrow Zn^{2+} \Rightarrow Cd^{2+}$ and $Zn^{2+} \Rightarrow Ni^{2+} \Rightarrow Cd^{2+}$). Differences in the variances calculated for each stage were similarly evaluated using an F-test (again considering both the order of metal ion exposure and the exposure history of the material).

Table 3 summarizes the resulting statistical comparisons of the native and chemically modified *D. innoxia* materials using the software package 'Analysis ToolPak' within Microsoft® Excel™. Mass balance values of the influent metal bound were used in these calculations. A confidence limit for each was 95% ($\alpha = 0.05$). The t-values presented indicate the respective probabilities that the amounts of the influent metal bound at each position were the same for both the native and modified materials. The results indicated that modification of the biomaterial did significantly decrease the binding capacities of the *D. innoxia* for each metal at each position in the sequential exposure sequence.

The F-values listed in Table 3 indicated the respective probabilities that the variances of the mean values between the native and modified biomaterials were statistically different. There were two cases a clear (>95%) difference in the variances was indicated; 1) the conditions when cadmium was the first metal exposed to the column and 2) when nickel was the third metal introduced to the biosorbent material. In both cases, larger variances were observed for cadmium binding to the chemically modified material, with a percent relative standard deviation (%RSD) of 27.22% compared to 4.86% for the native material. The binding of nickel showed similar behavior with a 25.97% RSD for its binding to the modified material in comparison to an RSD of 4.50% for the native biosorbent. This suggested sites involved in the binding of Cd^{2+} initially exposed to the material and those pertaining to Ni^{2+} binding as the third exposure metal ion were the least homogeneously affected by the chemical modification (i.e., esterification) reaction.

	Cd position 1	Cd position 2	Cd position 3
P T ≤ t (two tail)	1.01E-04	3.72E-05	3.18E-07
P F ≤ f (one tail)	0.024	0.258	0.261
	Ni position 1	Ni position 2	Ni position 3
P T ≤ t (two tail)	1.58E-03	2.73E-03	2.86E-03
P F ≤ f (one tail)	0.119	0.419	0.006
	Zn position 1	Zn position 2	Zn position 3
P T ≤ t (two tail)	3.14E-07	1.66E-06	8.53E-07
P F ≤ f (one tail)	0.390	0.484	0.492

Table 3. Comparing influent metal bound at each position for the native and modified biomaterials. The P T ≤ t value indicated the statistical probability that the mean values for the native and modified *D. innoxia* are the same. The P F ≤ f value indicated the probability that the variances of the two means are the same.

One question that arose with regard to the sequential exposure of the sorbent to chemically similar metal ions is whether there is a statistically significant difference in the binding capacities as a function of the position of the metal in the exposure sequence (i.e., first, second, or third). This is related to the possible presence of cooperativity in the formation (or elimination) of binding sites for one metal because of the earlier binding of another metal ion to a (presumably) nearby site.

Table 4 summarizes a comparison of influent metal binding capacities within each of the studies based on metal position in the sequence using a Student t-test at 95% confidence. The t-values listed suggest there is no statistical difference in the amount of influent metal bound for either the native or modified biomaterials based on the position of each metal in the exposure sequence. Conversely, the F-values suggested statistical differences in the variances based on position in the sequence for each of four cases: For nickel, binding to the native column comparing positions 1 and 3, and positions 2 and 3;. Also for cadmium exposed to the modified column comparing positions 1 and 3, and positions 2 and 3. These values again suggested greater inhomogeneity in the impact of the chemical modification procedure on the sites involved in the binding of these metal ions.

	1 vs. 2	1 vs. 3	2 vs. 3
T Cd native	0.739	0.671	0.985
T Ni native	0.196	0.462	0.338
T Zn native	0.380	0.707	0.596
T Cd modified	0.241	0.195	0.912
T Ni modified	0.699	0.686	0.955
T Zn modified	0.567	0.182	0.428
F Cd native	0.119	0.253	0.296
F Ni native	0.407	0.035	0.022
F Zn native	0.399	0.485	0.413
F Cd modified	0.400	0.025	0.041
F Ni modified	0.223	0.382	0.320
F Zn modified	0.476	0.397	0.421

Table 4. Comparing influent metal bound at each position. T values are statistical probability that the mean metal bound at each position in sequence are the same. F values are statistical probability that their variances are the same.

Tables 5a-c provide a closer look at metal position by considering the history of metal exposure as well as position of each metal ion in the sequence. The t-values for all three metals indicated no statistical difference as a result of metal exposure history to the influent metal bound for either the native or modified biomaterial. The F-values again suggest statistical differences in the variances in Ni^{2+} binding based on exposure history. The modified biomaterial showed a difference in the variances for the comparison of the sequences CdZnNi with ZnCdNi, and CdNi with CdZnNi. Comparatively larger relative standard deviations (RSDs) were calculated for the sequences CdNi and ZnCdNi relative to that for the CdZnNi exposure sequence. A probable difference in variances for the comparison of CdNi with CdZnNi, and CdNi with ZnCdNi was also observed for the native material. Under these conditions, CdNi exhibited comparatively large standard deviation relative to those of the other two conditions.

3.3.3 Binding site matrix analysis

Traditional statistical analysis of the metal ion binding data suggested that the chemical modification of the *D. innoxia* material decreased the number of binding sites significantly, thus reducing metal binding capacities for the esterified biomaterial. Additionally, statistically significant changes in the Ni^{2+} binding variability suggested non-uniform changes in metal-specific sites that resulted from the esterification reaction. In an effort to extract more information about the binding behavior of the biomaterial towards these three metals, a secondary method of data analysis was undertaken.

	2a vs. 2b	3a vs. 3b	2a vs. 3a	2a vs. 3b	3a vs. 2b	3b vs. 2b
T Native	0.955	0.471	0.607	0.710	0.830	0.784
F Native	0.105	0.384	0.345	0.247	0.183	0.264
T Modified	0.328	0.969	0.401	0.326	0.621	0.623
F Modified	0.268	0.196	0.416	0.148	0.207	0.060

Table 5a. Effect of column history on influent cadmium bound. T values are probability that the two compared means are statistically equivalent. F values are statistical probability that their variances are equivalent. $2a \rightarrow NiCd$, $2b \rightarrow ZnCd$, $3a \rightarrow NiZnCd$, $3b \rightarrow ZnNiCd$.

	2a vs. 2b	3a vs. 3b	2a vs. 3a	2a vs. 3b	3a vs. 2b	3b vs. 2b
T Native	0.483	0.104	0.326	0.557	0.345	0.665
F Native	0.093	0.454	0.036	0.043	0.267	0.304
T Modified	0.760	0.183	0.310	0.536	0.427	0.377
F Modified	0.459	0.034	0.047	0.413	0.055	0.374

Table 5b. Effect of column history on influent nickel bound. T values are probability that the two compared means are statistically equivalent. F values are statistical probability that their variances are equivalent. $2a \rightarrow CdNi$, $2b \rightarrow ZnNi$, $3a \rightarrow CdZnNi$, $3b \rightarrow ZnCdNi$.

	2a vs. 2b	3a vs. 3b	2a vs. 3a	2a vs. 3b	3a vs. 2b	3b vs. 2b
T Native	0.214	0.515	0.193	0.542	0.959	0.274
F Native	0.939	0.953	0.736	0.691	0.553	0.221
T Modified	0.412	0.327	0.383	0.440	0.470	0.354
F Modified	0.094	0.392	0.357	0.264	0.157	0.224

Table 5c. Effect of column history on influent zinc bound. T values are probability that the two compared means are statistically equivalent. F values are statistical probability that their variances are equivalent. $2a \rightarrow CdZn$, $2b \rightarrow NiZn$, $3a \rightarrow CdNiZn$, $3b \rightarrow NiCdZn$.

Three assumptions were made in pursuing this avenue of exploration. 1) There exist common binding sites shared between all three of the metal ions (δ_0). 2) For each metal ion, there exist ion-specific binding sites available only to that particular ion (α, β, γ). 3) For each stage in the influent solution exposure sequence, there may be enhancement or inhibition of binding due to the history of metal exposure, i.e., some level of cooperativity between sites. Making these assumptions allows for the construction of an overall binding equation for the three-metal system:

$$M_{Bound} = [\ \alpha_{Cd}\, X_{Cd} + \alpha_{Ni}\, X_{Ni} + \alpha_{Zn}\, X_{Zn}\ ] + \{\ \beta_{CdNi}\, X_{CdNi} + \beta_{CdZn}\, X_{CdZn}$$
$$+ \beta_{NiCd}\, X_{NiCd} + \beta_{NiZn}\, X_{NiZn} + \beta_{ZnCd}\, X_{ZnCd} + \beta_{ZnNi}\, X_{ZnNi}\ \} + (\ \gamma_{CdNiZn}\, X_{CdNiZn} + \tag{1}$$
$$\gamma_{CdZnNi}\, X_{CdZnNi} + \gamma_{NiCdZn}\, X_{NiCdZn} + \gamma_{NiZnCd}\, X_{NiZnCd} + \gamma_{ZnCdNi}\, X_{ZnCdNi} + \gamma_{ZnNiCd}\, X_{ZnNiCd}\ )$$
$$+ \delta_0\, X_{common}$$

From the sequential experiments, the amount of total metal bound is known for each of the above situations except for metal binding to the common binding sites, X_{Common}. While the site-type X_{Common} is available for each condition, the degree to which it is available may be limited by the metal exposure history on the column and the comparative affinities of each metal for those sites. To account for this, a secondary set of data was incorporated into the ensuing analysis. The total metal-bound data from the simultaneous exposure of all three metals to the biomaterial were indicative of situations in which all three unique metal binding sites (X_{Cd}, X_{Ni}, and X_{Zn}) are used along with only the common site, X_{Common}. This enabled the isolation of the common site from any binding enhancement or inhibition that could be attributed to metal history. By using the total amounts of each metal bound for each stage in the sequences and the amount of metal bound during the simultaneous exposure of the material to the three metals, 16 equations for metal binding can be written. The variables X_{Cd} through X_{common} can have values of either 1 (the site-type is involved) or 0 (the site-type is not involved). These can then be combined into a single, 16 by 16 matrix (X in Table 6) with the corresponding coefficients comprising the contents of 1 by 16 vector (c). For each stage in the sequences the common site, the individual metal-ion sites, and the corresponding sequential site will make contributions to the total metal bound. For example, the total metal bound at the sequence stage $Ni^{2+} \rightarrow Zn^{2+}$ can be represented by the equation:

$$M_{Total} = \alpha_{Ni} + \alpha_{Zn} + \beta_{NiZn} + \delta_0 \tag{2}$$

And the simultaneous exposure of all three metals can be represented by:

$$M_{Total} = \alpha_{Cd} + \alpha_{Ni} + \alpha_{Zn} + \delta_0 \tag{3}$$

Where M_{Total} is the total metal bound and the coefficients (α, β, γ, and δ) indicate the contribution of each site-type to the total metal bound.

	α Cd	α Ni	α Zn	β CdNi	β CdNi	β CdNi	β CdNi	β CdNi	β CdNi	γ CdNiZn	γ CdZnNi	γ NiCdZn	γ NiZnCd	γ ZnCdNi	γ ZnNiCd	δ_0
Cd	1	0	0	0	0	0	0	0	0	0	0	0	0	0	0	1
Ni	0	1	0	0	0	0	0	0	0	0	0	0	0	0	0	1
Zn	0	0	1	0	0	0	0	0	0	0	0	0	0	0	0	1
CdNi	1	1	0	1	0	0	0	0	0	0	0	0	0	0	0	1
CdZn	1	0	1	0	1	0	0	0	0	0	0	0	0	0	0	1
NiCd	1	1	0	0	0	1	0	0	0	0	0	0	0	0	0	1
NiZn	0	1	1	0	0	0	1	0	0	0	0	0	0	0	0	1
ZnCd	1	0	1	0	0	0	0	1	0	0	0	0	0	0	0	1
ZnNi	0	1	1	0	0	0	0	0	1	0	0	0	0	0	0	1
CdNiZn	1	1	1	0	0	0	0	0	0	1	0	0	0	0	0	1
CdZnNi	1	1	1	0	0	0	0	0	0	0	1	0	0	0	0	1
NiCdZn	1	1	1	0	0	0	0	0	0	0	0	1	0	0	0	1
NiZnCd	1	1	1	0	0	0	0	0	0	0	0	0	1	0	0	1
ZnCdNi	1	1	1	0	0	0	0	0	0	0	0	0	0	1	0	1
ZnNiCd	1	1	1	0	0	0	0	0	0	0	0	0	0	0	1	1
Simultaneous	1	1	1	0	0	0	0	0	0	0	0	0	0	0	0	1

Table 6. Matrix representing the contributions to total metal ion bound to the immobilized D. innoxia at each stage of influent metal ion exposure in the sequential experiments. A '1' indicates a contribution to total metal binding at the particular stage, a '0' indicates the site is not involved.

(A) Coefficient | Case	Exp. µmole/gram	1	2	3	4
Cd	25.53	29.58	62.68	187.30	-6.53
Ni	22.57	43.73	67.24	201.40	-1.94
Zn	25.40	53.4	65.29	211.10	0.36
CdNi	30.59	32.75	116.00	42.71	0.02
CdZn	31.03	47.59	112.80	57.55	1.28
NiCd	30.39	45.32	128.80	55.28	12.62
NiZn	29.94	47.78	113.60	57.74	-3.17
ZnCd	31.69	38.20	112.00	48.16	1.00
ZnNi	32.07	56.72	120.30	66.68	14.49
CdNiZn	31.27	31.87	167.60	-78.40	0.30
CdZnNi	28.80	32.57	162.60	-77.70	-0.96
NiCdZn	30.93	44.71	193.60	-65.50	13.17
NiZnCd	28.80	24.26	156.20	-86.00	-4.68
ZnCdNi	38.10	31.82	169.70	-78.40	2.56
ZnNiCd	28.69	30.96	167.30	-79.30	6.26
Simultaneous	41.72	36.41	191.70	410.20	-30.40

(B) Coefficient | Case	Exp. µmole/gram	1	2	3	4
Cd	50.35	45.60	756.48	334.45	-16.11
Ni	49.42	38.00	268.98	326.85	-11.68
Zn	52.27	46.14	-202.78	334.99	-7.64
CdNi	70.31	50.81	557.30	90.26	-8.63
CdZn	69.39	55.57	82.16	95.02	-3.49
NiCd	61.10	46.83	553.32	86.28	-8.19
NiZn	67.60	50.40	76.99	89.85	6.78
ZnCd	59.34	54.47	560.96	93.92	3.00
ZnNi	69.16	52.05	78.64	91.50	8.20
CdNiZn	74.07	39.35	-102.89	-128.61	-4.86
CdZnNi	73.12	40.19	-102.05	-127.77	0.24
NiCdZn	67.28	37.77	-104.47	-130.19	-2.02
NiZnCd	66.05	37.36	-104.88	-130.60	1.84
ZnCdNi	64.08	40.10	-102.14	-127.86	4.35
ZnNiCd	68.13	39.51	-102.73	-128.45	3.77
Simultaneous	72.87	40.68	-1505.70	673.19	-50.24

Table 7. Modified (A) and Native (B) columns experimental and calculated total metals bound, Case 1: All unique single metal bonds are included when appropriate in the sequential case and only all three unique sites along with the common site for the simultaneous case. Case 2: Same as case 1 except for the simultaneous all three, three metal binding site types are active along with the unique sites and common site. Case 3: The unique metal coefficients are not included in the simultaneous portion of the matrix calculation. Case 4: The unique metal coefficients are only included when they are the first metal on the column

The system can then be represented by the matrix equation:

$$[M] = [c] [X] \qquad (4)$$

with $[A]$ is a 1 by 16 vector containing the total amount of metal bound for each situation, $[c]$ is a 1 by 16 vector containing the contribution coefficients for each site type, and $[X]$ is the 16 by 16 matrix describing the types of binding that may be taking place. For example, the row in the matrix $[X]$ corresponding to the previous example (Ni $\rightarrow$Zn) would be [0 1 1 0 0 0 1 0 0 0 0 0 0 0 0 1]. Both $[M]$ and $[X]$ are therefore known or determined experimentally. The contribution coefficient matrix, $[c]$, can then be calculated by solving the multivariate equation.

$$[M] [X]^T ([X] [X]^T)^{-1} = [c] \qquad (5)$$

The superscripts T and -1 designate the corresponding transposed and inverted matrices, respectively.

Four matrices were examined using this methodology. The matrix presented in Table 6 includes all of the unique single metal bonds, where appropriate, in the sequential case, and only the three unique sites plus the common site for the simultaneous case. Tables 7a and b list the experimentally determined total amounts of metal bound with the predicted amounts from four separate theoretical calculations. Case 1 shows the results from the calculations using the matrix shown in Table 6. Case 2 is the same matrix as case 1 except for the simultaneous row, which now includes all three, three-metal binding sites (coefficients γ_{NiZnCd}, γ_{ZnCdNi}, and γ_{ZnNiCd}) along with the unique and common sites. Case 3 differs from case 2 by eliminating the unique sites from the simultaneous portion of the matrix. Case 4 is distinguished by only including the unique metal site type when the metal is the first metal introduced to the column.

Upon examining the results presented in Tables 7 a and b, it was evident that case 1 best approximated the experimental results. Figure 5 illustrates the values of the contribution coefficients for both the native and the modified (shaded) *D. innoxia* total metal bound studies.

The most striking and least surprising result of this analysis was the contribution the common site (δ_0) made to overall binding for both the native and the modified biomaterial. Also noteworthy, was the series of positive coefficients present in the two metal ion systems. For both the native and modified biomaterials, it appears that the presence of a metal ion enhanced the biomaterial's binding capacity. The lone exception to this was the impact of the NiZn sequence on the modified material.

The native biomaterial exhibited slight positive coefficients for the unique Cd^{2+} and Zn^{2+} sites and a moderate apparent inhibition for Ni^{2+}. All binary combinations of metal ions exposed to the native biomass resulted in moderate positive coefficient values. The tertiary combinations all yielded moderately negative values. This does not necessarily indicate an absolute inhibition of binding, but can be interpreted in terms of relative inhibition effects. Review of the experimental data listed in Table 7 b reveals single metal values as all near 50 μmol g^{-1} (average 50.68). Comparatively, binary combinations ranged from 60 – 70 μmol g^{-1} (average 66.15), an increase of 15 while tertiary combinations ranged from 65 –75 μmol g^{-1} (average 68.78), an increase which may not be statistically significant. This suggests some degree of cooperativity in metal-ion binding while the primary mechanism of metal ion binding is simple electrostatic (i.e., the dominance of the common sites).

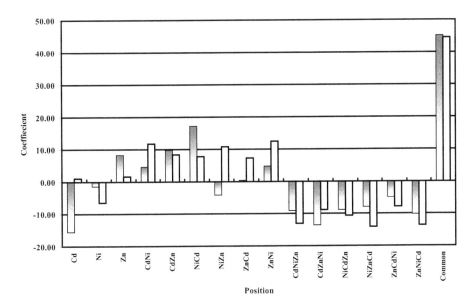

Fig. 5. Comparison of native and modified (shaded) *D. innoxia* matrix coefficients.

Binding Sites	Coefficient	Native Material	Modified Material
Cd	α_{Cd}	1.07	-15.57
Ni	α_{Ni}	-6.53	-1.42
Zn	α_{Zn}	1.61	8.25
CdNi	β_{CdNi}	11.74	4.59
CdZn	β_{CdNi}	8.36	9.76
NiCd	β_{CdNi}	7.76	17.16
NiZn	β_{CdNi}	10.79	-4.20
ZnCd	β_{CdNi}	7.26	0.37
ZnNi	β_{CdNi}	12.44	4.74
CdNiZn	γ_{CdNiZn}	-13.07	-9.13
CdZnNi	γ_{CdZnNi}	-8.85	-13.60
NiCdZn	γ_{NiCdZn}	-10.67	-8.86
NiZnCd	γ_{NiZnCd}	-14.11	-7.95
ZnCdNi	γ_{ZnCdNi}	-7.84	-4.96
ZnNiCd	γ_{CdNiZn}	-13.61	-10.19
Common	δ_0	44.53	45.15

Table 8. Coefficient values for both the native and modified D. innoxia metal ion binding sequences.

The modified biomaterial's coefficients behaved in a similar manner as the native. The single metal coefficients for Cd^{2+} and Ni^{2+} were both negative (Ni^{2+} slightly, Cd^{2+} moderately) while Zn²⁺ was moderately positive. All of the binary combinations were moderately positive except for NiZn, which was moderately negative, and ZnCd, which was only slightly positive. The tertiary combinations were as on the native biomaterial, all moderately negative. Again, this only reflects relative inhibition of metal ion binding. The data listed in Table 7a reveals single metal bound values were all about 25 μmol g⁻¹ (average 24.5). These are compared to those for the binary combinations which ranged from 30 – 32 μmol g⁻¹ (average 31.0), an increase of 6. Similarly to the native material, the tertiary combinations ranged from 29 – 39 μmol g⁻¹ (average 31.1), a slight increase which may not be statistically significant. This again suggested that some degree of cooperativity in binding with the primary binding mechanism involving electrostatic attractions.

4. Conclusions

Carboxylate groups are important functionalities for metal ion adsorption for the biomaterial *Datura innoxia*. Esterification of these functionalities was observed to reduce metal binding capacity 30-40%. Three metals of similar binding affinity to the biomaterial (Cd^{2+}, Ni^{2+}, and Zn^{2+}) were exposed sequentially to both the native esterified biomaterials immobilized within a polysilicate matrix. Removal of surface carboxylates significantly impacted the binding of each metal ion and the impact of each metal bound on the binding of the others. A model involving at least 16 binding sites for these metal ions revealed a majority of binding to involve species-independent or common sites and the presence of cooperativity for some metal-ion binding environments

5. Acknowledgements

The authors wish to acknowledge the financial support of the Waste-management Education and Research Consortium funded by the US Department of Energy through New Mexico State University.

6. References

Davis, T.A.; Volesky, B.; Mucci, (2003) A. A Review of the Biochemistry of Heavy Metal Biosorption by Brown Algae. *Water Research* vol.37, No. 18 (November 2003) pp. 4311-4330 ISSN: 0043-1354.

Drake, L.R.;.Rayson, G. D. (1996) Plant Materials for Metal Selective Binding and Preconcentration *Analytical Chemistry*, vol. *68*, no. 1 (Jauary 1996), pp. 22A-27A. ISSN 0003-2700

Drake, L.R.; Lin, S.; Rayson, G.D. (1996) Chemical Modification and Metal Binding Studies of *Datura innoxia. Environmental Science and Technology*, vol. *30*, no. 1 (January 1996) pp. 110-114. ISSN 0013 936X

Drake, L.R.; Hensman, C.E.; Lin, S.; Rayson, G.D.; Jackson, P.J. (1997) Characterization of Metal Ion Binding Sites on *Datura innoxia* by Using Lanthanide Ion Probe Spectroscopy. *Applied Spectroscopy*, vol. *51*, no. 10 (October 1997) pp. 1476-1483. ISSN 0003-7028

Fourest, E.; Volesky, B. (1996) Contribution of Sulfonate Groups and Alginate to Heavy Metal Biosroption by Dry Biomass of *Sargassum fluitans Environmental Science and Technology* vol. *30*, no. 1 (January 1996) pp. 277-282. ISSN 0013 936X

Gardea-Torresdey, J.L.; Tiemann, K.J.; Gamez, G.; Dokken, K. (1999) Effects of Chemical Competition for Multi-metal Binding by *Medicago sativa* (alfalfa), *Journal of Hazardous Materials,* vol. *69*, no. 1, (October 1999) pp. 41-51 ISSN: 0304-3894.

Gardea-Torresdey, J.L.; Arteaga, S.; Tiemann, K.J.; Chinaelli, R.; Pingatore, N.; Mackay, W., (2001) Absorption of Copper(II) by Creosote Bush (*Larrea tridentata*): Use of Atomic and X-ray Absorption Spectroscopy, *Environmental Toxology and Chemistry*, vol. *20*, no. 11 (November 2001) pp. 2572-2579 ISSN: 0730-7268.

Jackson, P.J.; Anderson, W.L;. DeWitt, J.G.; Ke, H.-Y.D.; Kuske, C.R.; Moncrief, R.M.; Rayson G.D., (1993) Accumulation of Toxic Metal Ions on Cell Walls of Datura innoxia Suspension Cell Cultures *In Vitro Cellular Developmental Biology-Plant*, vol. *29P*, no. 4 October 1993) pp. 220-226 ISSN: 1054-5476.

Kelley, C.; Mielke, R.E.; Dimaquibo, D.; Curtis, A.J.; DeWitt, J.G. (1999) Adsorption of Eu(III) onto Roots of Water Hyacinth. *Environmental Science and Technology*, vol. *33*, no. 9 (May 1 1999) pp. *1430-1443.* ISSN 0013 936X

Lin, S.; Drake, L.; Rayson, G. D. (1996) Applications of Frontal Affinity Chromatography to the Study of Interactions Between Metal Ions and a Complex Biomaterial *Analytical Chemistry*, vol. *68*, no. 23 (December 1 1996) pp. 4087-4093. ISSN 0003-2700

Lin, S.; Drake, L. R.; Rayson, G. D. (2002) Affinity Distributions of Lead Ion Binding to an Immobilized Biomaterial Derived from Cultured Cells of *Datura innoxia Advances in Environmental Research* vol. *6*, no. 4 (October 2002) pp. 523-532 ISSN: 1093-0191.

Majidi, V.; Laude, D.A.; Holcombe, J.A. (1990) Investigation of the Metal Algae Binding-Site with Cd-113 Nuclear-Magnetic-Resonance. *Environmental Science and Technology*, vol. *24*, no. 9 (September 1990) pp. 1309-1312. ISSN 0013 936X

Riddle, S.G.; Ran, H.H.; DeWeitt, J.G.; Andrews, J.C. (2002) Field, Laboratory, and X-ray Absorption Spectroscopic Studies of Mercury Accumulation by Water Hyacinths. *Environmental Sciency and Technology*, vol. *36*, no. 9 (May 1 2002) pp. 1965-1970. ISSN 0013 936X

Serna, D.D.; Moore, J.L.; Rayson, G.D. (2010) Site-Specific Binding Isotherms to a *Datura innoxia* Biosorbent. *Journal of Hazardous Materials*, vol. *173*, no. 1-3 (January 15 2010) pp.409-414 ISSN: 0304-3894.

Stark P. C.;. Rayson, G. D. (2000) Comparison of Metal Ion Binding to Immobilized Biogenic Materials in a Flowing System, *Advances in Environmental Research*, vol. *4*, no. 2 (May 2000) pp. 113-122 ISSN: 1093-0191.

Volesky, B. (2000) Biosorption of Heavy Metals: Methodology Example of Uranium Removal in *Biologische Abwasserreinigung*, vol. 14, no. , pp. 17-37.

Williams, P.A.; Rayson, G. D. (2003) Simultaneous Multi-element Detectino of Metal Ions Bound to a *Datura innoxia* Material, *Journal of Hazardous Materials* vol. B99, no. 3 (May 30 2003) pp. 277-285 ISSN: 0304-3894.

Xia, H.; Rayson, G. D. (1998) Investigation of Al binding to a *Datura innoxia* material using ^{27}Al NMR *Environmental Science and Technology*, vol. 32, no. 18 (September 15 1998) pp. 2688-2692. ISSN 0013 936X

Xia, H.; Rayson, G.D., (2002) [113]Cd NMR Spectrometry Of Cd Binding Sites On Algae And Higher Plant Tissues, *Advances in Environmental Research*, vol. 7, no. 1 (November 2001) pp. 157-167 ISSN: 1093-0191..

8

Research on Mg-Zn-Ca Alloy as Degradable Biomaterial

B.P. Zhang[1,2], Y. Wang[2] and L. Geng[2]
[1]National Engineering Laboratory for Carbon Fiber Technology,
Institute of Coal Chemistry, Chinese Academy of Sciences
[2]School of Materials Science and Engineering, Harbin Institute of Technology
China

1. Introduction

Magnesium and magnesium alloys are light metals, which characterized a low density, high specific strength and strong specific stiffness. The fracture toughness of magnesium is greater than that of ceramic biomaterials such as hydroxyapatite. The Young's elastic modulus and compressive yield strength of magnesium are closer to those of cortical bone. Especially, Mg^{2+} is present in large amount in the human body and involved in many metabolic reactions and biological mechanisms. The human body usually contains approximately 35g per 70kg body weight and the human body's daily demand for Mg is about 350 mg/day. Due to the excellent biomechanical properties and biocompatibility, magnesium alloys used to be introduced as implants into orthopedic and trauma surgery in recently years [1~3].Various magnesium alloys have been investigated as biodegradable materials and some of them have been shown good biocompatibility. For example, AZ31, AZ91, WE43, LAE442, Mg-Ca and Mg-Zn have been investigated for bone implant application [4~8]. It has been shown that magnesium enhances osteogenesis response and increases newly formed bone. However, some magnesium alloys containing aluminum or heavy metal elements which have latent toxic effects on the human body. Thus, several problems such as inadequate strength, rapid corrosion and toxic ions must be solved before this unique metal is widely used in biomedical fields.

It is well known that pure magnesium has poor mechanical properties and the mechanical properties of magnesium can be effectively improved by the appropriate selection of alloying elements [1]. But, based on the aforementioned considerations, the range of alloying elements used in the degradable magnesium alloys is rather limited, Zn, Mn, Ca and perhaps a very small amount of low toxicity RE can be tolerated in the human body and can also be retard the biodegradation. Therefore, Mg-Ca binary alloys attract attention of researchers because Ca is an important element of human bones. The mechanical properties and biocompatibility of Mg-Ca binary alloy can be adjusted by controlling the Ca content and processing treatment. However, an inadequate mechanical properties as well as lower corrosion resistances of Mg-Ca binary alloys are the biggest drawback of these alloys [7][8]. Fortunately, in latest recent years, Mg-Zn system is paid more attention because Zn is one of abundant nutritional elements in human body [9] [10]. Additionally, it is a great potential alloying element to improve the mechanical properties and corrosion resistance of Mg alloys [11][12]. And the addition of other alloying element can further improve the mechanical properties of Mg-Zn alloys [13] [14]. Zn/Mn-containing magnesium alloys, e.g. Mg2Zn0.2Mn [15] and Mg-1.2 Mn-1.0 Zn [16] ternary alloys are studied, the results indicate that Zn/Mn-containing magnesium alloys have satisfactory mechanical properties and can be potential biodegradable alloys. But, the degradation rates of Zn/Mn-containing magnesium alloys are so fast. After 9 weeks implantation, about 10~17% Mg-Mn-Zn magnesium implant has degraded. After 18 weeks implantation, about 54% Mg-Mn-Zn alloy has degraded [16]. The results studied by H.X. Wang at al [17] indicate that the Mg-Zn-Ca alloys coated with Ca-deficient hydroxyapatite have an excellent corrosion resistance in Kokubo's simulated body fluid (SBF), but the chemical composition of Mg-Zn-Ca alloys was not reported. L.Mao et al [18] studied the effects of Zn on microstructure and mechanical properties of biomedical Mg-Ca-Zn alloys. The results show that the microstructure is refined and the mechanical properties can be improved evidently with Zn content increasing. The mechanical properties of bending and compression can meet the requirements for hard tissue metal implants. However, the effect of Ca on microstructure and mechanical properties of biomedical Mg-Ca-Zn alloys, the corrosion resistance and cytotoxicity were not studied. Xuenan Gu et al[19] reported that the Mg66Zn30Ca4 bulk metallic glasses sample presents a more uniform corrosion morphology than as-rolled pure Mg and Mg70Zn25Ca5 samples. Both indirect cytotoxicity and direct cell culture experiments were carried out using L929 and MG63 cell lines. The results show higher cell viabilities for Mg-Zn-Ca extracts than that for as-rolled pure Mg. In addition, L929 and MG63 cells were found to adhere and proliferate on the surface of Mg66Zn30Ca4 sample. Unfortunately，the cytotoxicity was tests by MTT, according Janine Fischer et.al[20] research, in the case of Mg materials, the use of MTT test kits leads to false positive or false negative results, because Mg is a very reactive element. It is conceivable that Mg in the highly alkaline environment may be able to open the ring form of the tetrazolium salt and bind to it, which could lead to a change in colors similar to the formation of formazan in the case of the MTT tests with cells.

It is reasonable to speculate that the Mg-Zn-Ca alloys with a proper Zn and Ca content can exhibit a superior combination of mechanical properties, corrosion resistance and biocompatibility. In this paper, Zn and Ca, which have no toxicity, are chosen as alloying elements to successfully improve the mechanical properties of magnesium. The effects of Zn and Ca content on mechanical properties, in-vitro corrodible property and cytotoxicity of Mg-Zn-Ca alloys have been systematic investigated to assess the feasibility of Mg-Zn-Ca alloys for use as bone implant materials.

2. Materials and methods

2.1 Materials

Mg-Zn-Ca alloys were prepared from high purity Mg (99.99 %), Purity Zn (99.8 %), and an Mg-26.9 wt. % Ca master alloy. Melting and alloying operations were carried out in a steel crucible under the protection of a mixed gas atmosphere of SF_6 (0.3 vol. %) and CO_2 (Bal.). Purity Zn and master alloy were added into the pure Mg melts at 720 °C. The melts were kept for 10min at 720°C to ensure that all the required alloying elements were dissolved in the melt alloy, and then the melts were cooled down to 700 °C and poured into a steel mold which had been pre-heated to 200 °C.

2.2 Composition and microstructure characterization

X-ray diffraction (XRD, Philips-X'Pert) using Cu Ka radiation was employed for the identification of the constituent phases in the as-cast Mg-Zn-Ca alloy and their corrosion products after immersion. Microstructure observations of the alloys were conducted on Olympus optical microscope. The specimens for optical microscopy were etched with a solution of 8 vol. % acetic acid for 30 s, thoroughly flushed with water and alcohol, and then dried by hot air.

2.3 Mechanical properties

Tensile tests were carried out using an Instron-5569 universal testing machine at a constant crosshead speed of 1.0mm/min at room temperature. The tensile specimens with diameter of 6mm and gauge length of 30mm were cut by electric-discharge machining from the ingot. The Young's modulus was get from the tensile test. The fracture morphologies were examined by SEM (SEM, Hitachi-3000N).

2.4 In vitro degradation tests

Electro-chemical measurements and immersion tests were performed in a Hank's simulated body fluid to evaluate the in-vitro degradation properties. The chemical composite of Hank's simulated body was listed in table 2.The pH value of Hank's solution was adjusted with HCl and NaOH to 7.2~7.4, to avoid precipitation or formation of sediments in the solution. Its temperature was controlled around 37 ± 0.5 ^{0}C, which is equal to the human body normal temperature. Pure Mg (>99.99%) was also tested as a contrast.

2.4.1 Electrochemical measurements

A typical three-electrode system consisting of graphite rod as counter electrode, saturated calomel electrode (SCE) as a reference electrode and specimen ($1cm^2$ exposed areas) as a working electrode was used. Potentiodynamic polarization experiments were carried out at a scan rate of 0.5 mV/s. The electrochemical measurements of specimens with thickness of 4 mm and a gauge diameter of 15mm were machined from the ingot and ground with 2000 grit SiC paper, and they were rinsed with distilled water and dried by hot air.

2.4.2 Immersion tests

The immersion tests were carried out in Hank's solution according to ASTM-G31-72 [21]. Samples were removed after 30 days of immersion, rinsed with distilled water, and were cleaned with chromic acid to remove the corrosion products. The degradation rates (in units of mm year-) were obtained according to ASTM-G31-72. An average of five measurements was taken for each group. The pH value of the solution was also recorded in the immersion tests at absolute group for 144 hours.

2.5 Cytotoxicity assessments

L-929 cells were adopted to evaluate the cytotoxicity of Mg-Zn-Ca alloys. The cells were cultured in Dulbecco's modified Eagle's medium (DMEM), 10% fetal bovine serum (FBS), 100 Uml-1penicillin and 100 mg ml-1 streptomycin at 37 ^{o}C in a humidified atmosphere of 5% CO2. The cytotoxicity tests were carried out by indirect contact. Extracts were prepared using DMEM serum free medium as the extraction medium with the surface area of extraction medium ratio 1.25 ml/cm^2 in a humidified atmosphere with 5% CO_2 at 37 ^{o}C for

72 h. The supernatant fluid was withdrawn and centrifuged to prepare the extraction medium, then refrigerated at 4 °C before the cytotoxicity test. The control groups involved the use of DMEM medium as negative controls. Cells were incubated in 96-well cell culture plates at 5×10^4 cells/ml medium in each well and incubated for 24 h to allow attachment. The medium was then replaced with 100µl of extracts. After incubating the cells in a humidified atmosphere with 5% $CO2$ at 37 °C for 2, 4 and 7 days, respectively, cell morphology was observed by optical microscopy (Nikon ELWD 0.3 inverted microscope).The neutral red viability assay was performed according to published procedures. A stock solution of neutral red (Beyotime, China) was prepared in water (1%). The stock solution was diluted to 50 µg/ml in complete culture medium and 200µl of the staining solution were added to each well after removal of the exposure medium. The cells were incubated for 3 h at 37°C, The cells were then fixed with 200µl for-maldehyde/$CaCl2$(3.7%/1%) and destained with 200µl methanol/glacial acetic acid (50%/1%), The plates were shaken for 60 min at room temperature using a plate shaker. Optical densities were measured at 540 nm in a multiwell spectrophotometer (Bio-RAD 680).The cell relative growth rate (RGR) was calculated according to the following formula:

$$RGR= OD_{test} / OD_{negative} \times 100\%$$

2.6 Animal test
2.6.1 Surgery
Animal tests were approved by the Ethnics Committee of the First Affiliated Hospital of Harbin Medical University. The in-vivo degradation experiments were performed in the animal laboratory of the hospital. A total of 15 adult New Zealand rabbits (6 females), 2.0~2.5kg in weight, were used. In the experimental group, sodium pentobarbital (30mg kg-1) was administered to perform anesthesia by intravenous injection. The sterile Mg-Zn-Ca alloy rod sample was implanted into the femora of the rabbit.

After operation, all animals received a subcutaneous injection of penicillin to avoid a wound contamination and were allowed to move freely in their cages without external support. After operation, five rabbits were sacrificed randomly at 1, 2 and 3 months, respectively.

2.6.2 Degradation and histological analysis
The bone samples with magnesium implants were fixed in 2.5% glutaraldehyde solution and then embedded in epoxy resin for microstructure analysis. The samples were sliced by hard tissue slicer (ZJXL-ZY-200814-1). Samples were made perpendicular to the long axis of the implant to get a cross-section of the implant and surrounding bone tissue. The cross-section microstructure was observed by an optical microscope (Nikon ELWD 0.3 inverted microscope) and a scanning electronic microscope (Hitachi S-5500). The residual implant areas were measured on the cross-section images using analysis software. The ratio of the residual cross-section area of implants to the original cross-section area (residual area/implant area×100%) was used to assess the in vivo degradation rate of magnesium alloys. The element distributions in the residual implants and the degradation layer after 3 months implantation were analyzed.

For histological analysis, the bone samples with magnesium implants were fixed in 4% formaldehyde solution, dehydrated, and then decalcified in ethylene diamine tetra acetate. Then, the specimens were embedded in paraffin and cut into films with 5µm in thickness.

The films were then stained with Hematoxylin and eosin. Histological images were observed on an optical microscope.

2.7 Statistical analysis
A t-test was used to determine whether any significant differences existed between the mean values of the cytotoxicity and animal tests of the experiment. The statistical significance was defined as $P < 0.05$.

3. Results and discussion
3.1 Phase compositions and microstructures evolution of the as-cast Mg-Zn-Ca alloys
3.1.1 The effects of Zn content on phase compositions and microstructures of the as-cast alloys
In this study, in order to investigated the effects of Zn and Ca on the phase compositions and microstructures evolution of the as-cast Mg-Zn-Ca alloys, respectively, the initial content of Ca design as 0 wt. % and then changed the content of Zn to study the effects of Zn on phase compositions and microstructures. The chemical compositions of the Mg-xZn alloy obtained by ICP-AES were listed in Table 1. The impurity contents of the Mg-x Zn alloy were very low for better degradation properties and biocompatibility. X-ray diffraction (XRD) analyses were used to investigate the existing intermetallic phases in the Mg-x Zn Ca alloys (Fig. 1). As shown in Fig. 1, there was only α-Mg diffraction peaks phase in the Mg-1.0Zn alloy. Diffraction peaks from the Mg_2Zn phase was not detected. With the Zn concentration increasing, MgZn phase's patterns were began to detect in Mg-5.0 Zn and Mg-6.0 Zn alloy.

Materials	Chemical composition (wt.%)					
	Al	Zn	Mn	Si	Fe	Mg
Mg-1.0Zn	0.023	0.976	0.058	0.031	0.004	Balance
Mg-2.0Zn	0.033	1.852	0.030	0.039	0.007	Balance
Mg-3.0Zn	0.029	2.732	0.022	0.036	0.007	Balance
Mg-4.0Zn	0.019	3.925	0.021	0.032	0.008	Balance
Mg-5.0Zn	0.027	5.223	0.031	0.034	0.009	Balance
Mg-6.0Zn	0.024	5.977	0.019	0.033	0.012	Balance

Table 1. Chemical compositions of the as-cast Mg-Zn alloy

The microstructures of the as-cast Mg-x Zn alloys were shown in Fig.2. Fig. 2(a) was taken from Mg-1.0 Zn alloy, in which the microstructure consists of the α-Mg . The maximum solubility of Zn in the magnesium was about 2 wt. % at room temperature in the equilibrium state, when no more than 2 wt. % Zn was added, the Zn was solid solution in Mg matrix. When the contents of Zn was more than 4 wt. % , the microstructure obviously changed, there were more second phases precipitated and the morphogenesis of second phases were small particle. As shown in Fig.2 (f), with the increasing of Zn content, lamellar eutectic appears in the as-cast microstructure. The eutectic structures were very coarse and mostly distributed in the grain boundary and less in the areas of inter-dendrite,

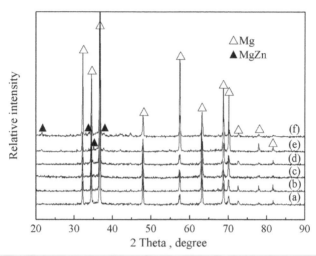

Fig. 1. XRD patterns of as-cast Mg-Zn alloys (a)Mg-1.0Zn; (b)Mg-2.0Zn; (c)Mg-3.0Zn; (d)Mg-4.0Zn; (e)Mg-5.0Zn; (f)Mg-6.0Zn

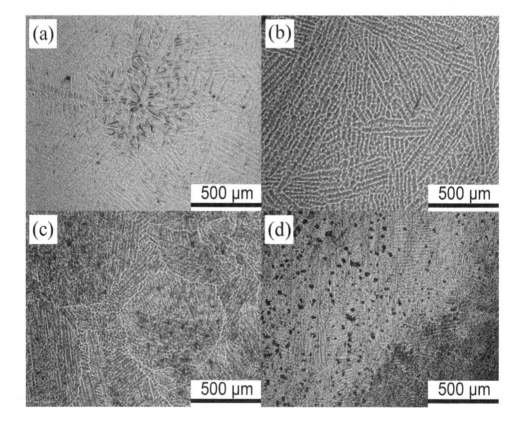

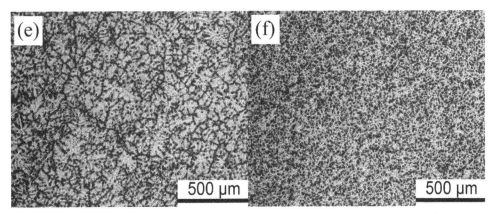

Fig. 2. Optical microstructure of as-cast Mg-Zn alloys (a) Mg-1.0Zn; (b) Mg-2.0Zn; (c) Mg-3.0Zn; (d) Mg-4.0Zn; (e) Mg-5.0Zn; (f) Mg-6.0Zn

3.1.2 The effects of Ca content on phase compositions and microstructures of the as-cast alloys

In present study, the mechanical properties show that when Zn content is 4wt. %, the Mg-xZn has good mechanical properties. Thus, the initial content of Zn designs as 4.0wt. % and then changed the content of Ca to study the effects on phase compositions and microstructures .The chemical compositions of the Mg-4.0 Zn-x Ca alloy obtained by ICP-AES were listed in Table 3. X-ray diffraction (XRD) analyses were used to investigate the existing phases in the Mg-4.0Zn-xCa alloys, and the results were shown in Fig.3. The results showed that α-Mg and MgZn phases were detected in the Mg-4.0 Zn alloy, and it also indicated that the diffraction peaks from the MgZn phase were very weak and that of the Mg were strong. There was no obvious change in diffraction peak when 0.2 wt. % Ca and 0.5wt. % Ca was added into the Mg-4.0 Zn alloy. With the Ca concentration increased to 1.5 wt. %, $Ca_2Mg_6Zn_3$ phases began to be detected in Mg-4.0 Zn-xCa. When the Ca concentration increased to 2.0 wt. %, Mg_2Ca, and $Ca_2Mg_5Zn_{13}$ phases began to be detected in the alloy.

Materials	Chemical composition (wt.%)						
	Al	Zn	Mn	Si	Fe	Ca	Mg
Mg-4.0Zn	0.023	3.926	0.058	0.031	0.004	0.007	Balance
Mg-4.0Zn-0.2Ca	0.033	1.852	0.030	0.039	0.007	0.180	Balance
Mg-4.0Zn-0.5Ca	0.029	2.732	0.022	0.036	0.007	0.452	Balance
Mg-4.0Zn-1.0Ca	0.019	3.925	0.021	0.032	0.008	0.915	Balance
Mg-4.0Zn-1.5Ca	0.027	5.223	0.031	0.034	0.009	1.635	Balance
Mg-4.0Zn-2.0Ca	0.024	5.977	0.019	0.033	0.012	2.158	Balance

Table 2. Chemical composition of the as-cast Mg-4.0Zn-xCa alloys

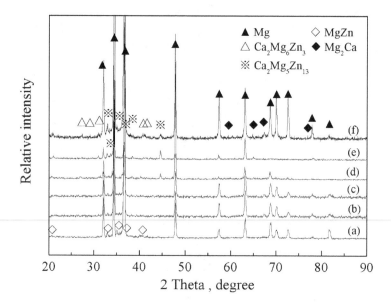

Fig. 3. XRD patterns of as-cast Mg-4.0Zn-xCa alloys (a) Mg-4.0Zn; (b) Mg-4.0Zn-0.2Ca; (c) Mg-4.0Zn-0.5Ca; (d) Mg-4.0Zn-1.0Ca; (e) Mg-4.0Zn-1.5Ca; (f) Mg-4.0Zn-2.0Ca

The microstructures of the as-cast Mg-4.0 Zn-x Ca alloys were shown in Fig.4. Fig. 4(a) was taken from the Mg-4.0 Zn alloy, which consisted of the dendrite α-Mg matrix and some polygonal shaped second phases which distributed in the areas of inter-dendrite and grain boundary. The second phases were very coarse in the Mg-4.0 Zn. Fig. 4(b) was taken from the Mg-4.0 Zn-0.2 Ca alloy, which indicated that the microstructure had an evidently change compared with Mg-4.0 Zn alloy, and the second phase changed its shape and distributed in the areas of inter grain. With the increase of Ca concentration, however, lamellar eutectic appeared in the as-cast microstructure, eutectic structure was mostly distributed in the grain boundary and little in the areas of inter-dendrite, as shown in Fig. 3 (e) and (f) which were taken from Mg-4.0 Zn-1.5 Ca and Mg-4.0 Zn-2.0 Ca alloys, respectively. It's easy to fond out that the morphogenesis of second phases have an obviously change by an increase in Ca content. At first the second phase was polygonal particles in Mg-4.0 Zn alloy, and then when less than 0.5 wt. % Ca was added in Mg-4.0 Zn alloy, the second phase changed its morphology, and it was small round particle. Finally when more than 0.5wt. % Ca was added, the second phase was lamellar structure.

In the initial stages of solidification, Zn and Ca were complete melts in the magnesium. Subsequently, as the solidification develops, the solute atoms are rejected by the growing α-Mg and enriched in the residual liquid, which began to form clusters precipitation in the grain boundary and inter dendrite arm space. When the Ca concentration was increasing to 1.5 wt. %, it was apt to forming lamellar eutectic.

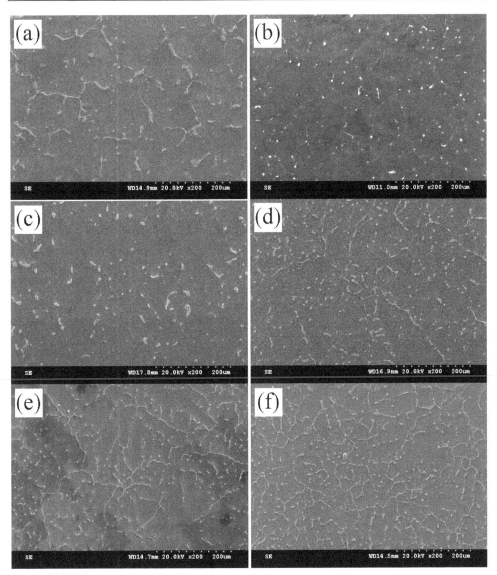

Fig. 4. Microstructure of as-cast Mg-4.0Zn-xCa alloys (a) Mg-4.0Zn; (b) Mg-4.0Zn-0.2Ca; (c) Mg-4.0Zn-0.5Ca; (d) Mg-4.0Zn-1.0Ca; (e) Mg-4.0Zn-1.5Ca; (f) Mg-4.0Zn-2.0Ca

3.2 Mechanical property evolution of the as-cast Mg- Zn-Ca alloys
3.2.1 The effects of Zn content on mechanical property of as-cast alloys
The tensile tests were carried out at room temperature on the as-cast Mg-x Zn alloys. The tensile strength and elongation of present alloy were shown in Table 3. The typical stress-train curves of Mg-x Zn alloys were depicted in Fig.5. As shown in Table 3 and Fig.5, after 1.0 wt.% Zn was added to the pure Mg, the mechanical properties of as-cast Mg-1.0Zn like

pure Mg was still weak, in which the yield strength was 21MPa, UTS was 101MPa and the elongation was 6.9%. With the increasing of Zn contents, the yield strength, UTS and the elongation was increased. When the Zn content was up to 4.0wt.%, the mechanical properties reach to the peak value, the yield strength was 58.1MPa, UTS was 216.85MPa and the elongation was 15.8%. The mechanical properties of Mg-Zn binary alloy when ulteriorly increasing of Zn contents was declined. Its UTS was 182 MPa, and its elongation was only 7.2 % for Mg-6.0 Zn alloy. Direct estimation of stacking fault energy by thermodynamic calculations showed that Zn reduces stacking fault energy of the Mg-Zn alloys. Stacking fault energy is an important physical properties of the material, which directly affects the mechanical properties, dislocation cross slip, phase stability and the dynamic recrystallization of metal materials. It has been confirmed that the stacking fault energy in magnesium alloy plays an important role in mechanical properties and the dynamic recrystallization[22]

Alloy	Yield strength(MPa)	UTS (MPa)	Elongation (%)
Mg-1.0Zn	20±2	101.5±3	6.96±0.5
Mg-2.0Zn	27±2	145.9±5	12.23±1.5
Mg-3.0Zn	47±1.5	167.8±10	13.7±1.0
Mg-4.0Zn	58±1.0	216.8±15	15.8±5.5
Mg-5.0Zn	68±1.5	185±5	9.2±0.5
Mg-6.0Zn	69±1.5	182±5	7.2±0.5

Table 3. Mechanical properties of Mg-xZn alloys at room temperature

3.2.2 The effects of Ca content on mechanical property of the as-cast alloys

The tensile strength and elongation of as-cast Mg-4.0 Zn-x Ca alloys were shown in table 4. The typical stress-train curves of as-cast Mg-4.0 Zn-x Ca alloys were depicted in Fig.6. The ultimate tensile strength (UTS) and elongation of as-cast Mg-4.0 Zn alloy were 180MPa and 9.5%, respectively. After 0.2 wt. % Ca was added, the UTS and elongation of as-cast Mg-4.0 Zn-0.2 Ca alloy were improved to 215 MPa and 17.5%, respectively. When 0.5 wt. % Ca was added, the Mg-4.0 Zn-0.5 Ca alloy has similar mechanical property as the Mg-4.0 Zn-0.2 Ca alloy. However, the mechanical properties of as-cast Mg-4.0 Zn-1.0 Ca alloy began to decline. When the Ca concentration was up to 2.0 wt. %, the alloy showed worse mechanical property, its UTS was 142 MPa and elongation was only 1.7 %.

Fig.7 showed the typical fracture surfaces of as-cast Mg-4.0 Zn-x Ca alloys. As it was showed that the fracture type was ductile fracture when Ca concentration was lower than 0.5 wt. %. Big dimples and tearing edges can be evidently observed on the fracture surface of the as-cast Mg-4.0 Zn-0.2 Ca (Fig.8 (a)).When the Ca concentration was 1.0 wt. %, the Mg-4.0 Zn-1.0 Ca alloy showed mixture fracture morphology. When the Ca concentration was up to 2.0 wt. %, the fracture type of the alloy was brittle fracture. The pearl-shaped fracture can be easily observed on the fracture surface of the as-cast Mg-4.0 Zn-2.0 Ca alloy (Fig. 5 (c)).

Alloy	Yield strength (MPa)	UTS (MPa)	Elongation (%)
Mg–4.0Zn	58±1.0	216.8±15	15.8±5.5
Mg–4.0Zn–0.2Ca	58.1±1.0	225±5	17.5±1.0
Mg–4.0Zn–0.5Ca	70±3.0	180±5	12.3±1.5
Mg–4.0Zn–1.0Ca	83±2.0	175±10	8.7±1.0
Mg–4.0Zn–1.5Ca	83±3.0	167±5	7.1±2.5
Mg–4.0Zn–2.0Ca	90±4.0	143±5	2.1±0.5

Table 4. Mechanical properties of Mg-4.0wt.%Zn-xCa alloys at room temperature

The mechanical properties of magnesium were affected by each alloying constituent. Zinc was an effective alloying ingredient in magnesium. Because zinc had a relatively high solid solubility in magnesium at high temperature, a good mechanical properties were achieved by solid solution strengthen. Binary Mg-Zn alloys like Mg-Al alloys, also respond to age hardening, and contrary to Mg-Al alloys, coherent GP zone and semeicoherent intermediate precipitate were formed to have an enhanced effect. However, in the Mg-Zn alloys, the maximum solubility of zinc in the magnesium drops to 1.6 wt. % (i.e. 0.6 at. %) at room temperature in the equilibrium state [23]. When the zinc content was more than 4.0 wt. %, in the solidification process, the melt zinc atoms would be rejected by the growing α-Mg and enriched in the residual liquid, these rich areas were often prone to formation of microporosity.

In Mg-Zn alloys, the progressive addition of Ca had been found to substantially increase the temperature difference between liquid and solid phase lines, which was conducive to the grain refinement in the solidification process. At the same time, the introduction of Ca to Mg-Zn alloys result in precipitation of desolventizing phase, Ca2Mg6Zn3 and Ca2Mg5Zn13, which could enhance the strength and toughness of alloy [24][25][26]. The current work showed that an addition of small amount of Ca to Mg-4.0 Zn alloys had a marked increase in the tensile strength, but Ca content was excess of 0.5 wt. % make the tensile strength prone to decrease. The precipitates in the Mg-4.0 Zn-0.2 Ca and Mg-4.0 Zn-0.5 Ca alloys were Ca2Mg6Zn3 and Ca2Mg5Zn13 phases, which were small particles in the alloys. Thus, the tensile property of Mg-4.0 Zn-0.2 Ca and Mg-4.0 Zn-0.5 Ca alloys were improved. However, the maximum solubility of Ca in the magnesium was only 0.2 wt. % at room temperature and 1.2 wt. % at high temperature in the equilibrium state, when more than 1.0 wt. % Ca was added, the precipitates in the grain boundary began to continuously precipitated and the morphogenesis of the precipitates were changed to lamellar structure, made the tensile properties decline. When the Ca concentration was up to 2.0 wt. %, in the grain boundary tends to form eutectic structure which caused the tensile property deteriorate.

3.3 In-vitro degradation tests
3.3.1 The effects of Zn content on in-vitro degradation of the as-cast alloys
The representative potentiodynamic polarization curves of Mg-xZn alloys in Hank's solution were shown in Fig.8, with pure Mg as contrast. As shown in Fig.8, the corrosion potential of the Mg-x Zn alloys was higher than that of pure Mg. The corrosion potential of

pure Mg was -1574mV. The corrosion potential was correlated with the Zn concentration. The corrosion potential of Mg-2.0Zn and Mg-3.0 Zn alloys were about -1561 and -1568 mV, respectively, which were nearly the same and about 10 mV high than that of pure Mg. The Mg-5.0 Zn and Mg-6.0 Zn alloy samples exhibit high corrosion potentials of about -1524 and -1547 mV, respectively, which were about 50 mV higher than that of pure Mg. It could be seen that the addition of Zn improved the corrosion potential of the as-cast Mg-x Zn alloys. But, the addition of elements Zn was also increased the current densities of the resulted as-cast Mg alloys in Hank's solution.

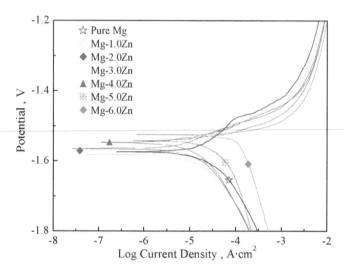

Fig. 8. Potentiodynamic polarization curves of Mg-Zn alloys in SBF solution

The reason for the increase corrosion potential of Mg-x Zn alloys was that the Zn element had a high electronegative. But, when the Zn concentration increased, the corrosion resistance was decreased. The reason for the reduced corrosion resistance of Mg-x Zn alloys was that the second phase precipitated during the solid solidification processes, which accelerated the corrosion rate due to the different electrochemical behaviors of α-Mg and precipitates.

Generally, the cathodic polarization curves were assumed to represent the cathodic hydrogen evolution through water reduction, while the anodic polarization curves represented the dissolution of magnesium. It could be seen that the cathodic polarization current of hydrogen evolution reaction on Mg-1.0 Zn alloy sample was much lower than that of Mg-5.0 Zn and Mg-6.0 Zn Ca alloys sample, suggesting that over potential of the cathodic hydrogen evolution reaction was lower for Mg-1.0 Zn and Mg-2.0 Zn alloys sample. As a result, the cathodic reaction was kinetically more difficult on the Mg-1.0 Zn alloy and Mg-2.0 Zn alloy sample than that on the Mg-5.0 Zn Ca alloy samples. The degradation rates of Mg-1.0Zn degraded were slower thanMg-5.0Zn, Mg-6.0Zn, which was adherence to the electrochemical results.

3.3.2 The effects of Ca content on in-vitro degradation of the as-cast Mg-Zn-Ca alloys

The representative potentiodynamic polarization curves of the pure Mg and Mg-4.0 Zn-x Ca alloys in Hank's solution were shown in Fig.9. The mean corrosion potentials of Mg-4.0Zn-xCa alloys are enhanced to -1574mV, which is increased by 70 mV compared with -1646 mV of the pure Mg corrosion potential. However, it is confirmed that Mg-4.0 Zn-0.2 Ca alloy exhibits the best corrosion resistance among Mg-4.0Zn-xCa alloys, even higher than that of Mg-4.0 Zn alloy through further observation. This particular phenomenon can be explained as follows. Firstly, the addition of 4.0 wt. % Zn can cause the formation of coarse MgZn precipitate as shown in Fig.4 and 2(a), which reduces the corrosion resistance of Mg-4.0Zn alloy due to the different electrochemical behaviors between primary α-Mg and precipitate. Then, the slight addition (less than 0.5 wt. %) of Ca alloying element can cause MgZn precipitates to be effectively transform to fine ternary precipitates, which has been clearly documented in previous literature [27][28]. The refinement and homogenization of precipitate phase can improve the corrosion resistance of Mg-4.0Zn-0.2Ca alloy compared with that of Mg-4.0Zn alloy. Finally, increasing Ca content was over than 0.5wt. % cause the formation of another coarse Mg_2Ca, $Ca_2Mg_6Zn_3$ and $Ca_2Mg_5Zn_{13}$ precipitates as shown in Fig.3 and 4. It is quite obvious that the precipitates increases with Ca content increasing, which decreases the corrosion resistance of as-cast Mg-4.0Zn-xCa alloys.

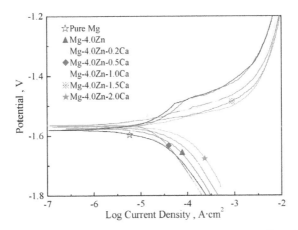

Fig. 9. The potentiodynamic polarization curves of as-cast Mg-4.0 Zn-x Ca alloys in Hank's solution.

It is well known that the cathodic polarization curves represent the cathodic hydrogen evolution through water reduction, while the anodic polarization curves do the dissolution of magnesium. That is to say, it is equivalent to that Mg-4.0 Zn-0.2 Ca alloy sample exhibits the lowest current of hydrogen evolution reaction and Mg-4.0 Zn-1.5 Ca and Mg-4.0Zn-2.0 Ca alloys samples does the highest ones, which indicates that over potential of the cathodic hydrogen evolution reaction of Mg-4.0 Zn-0.2 Ca alloy is much lower than those of Mg-4.0 Zn-1.5 Ca and Mg-4.0 Zn-2.0 Ca alloys. Therefore, the lowest cathodic hydrogen evolution reaction brings the highest corrosion resistance to the Mg-4.0 Zn-0.2 Ca alloy.

Fig.10 illustrates the pH variation of Hank's solution versus the immersion testing time for Mg-4.0Zn-xCa alloys. It could be observed that the pH variations of the alloys all obey the parabolic rate law. The pH variation rate decreases with the immersion time increasing. After 48 hrs immersion, all the pH values of the samples tend to be stable. In the early period of immersion, both pure Mg and the Mg-4.0Zn-xCa alloys acutely reacted with

Hank's solutions and rapidly generated bubbles. And these reactions of Mg and H$_2$O in Hank's solution generated a large amount of OH- and leaded to the pH values of the solutions be obviously increased. Comparing the pH values of Mg-4.0Zn-xCa alloys, it can be found that the pH variations of pure Mg and Mg-4.0Zn-0.2 Ca alloy are much lower than those of Mg-4.0Zn-1.5Ca and Mg-4.0Zn-2.0Ca alloys. The pH values of Mg-4.0Zn-1.5Ca and Mg-4.0Zn-2.0Ca alloys are remarkably increased to 8.2 from 7.4 after 12hrs immersion tests, which is even equal to those of pure Mg and Mg-4.0Zn-0.2Ca alloy after 96hrs immersion tests. At the end of the immersion tests,the pH values are increased to 8.22 and 8.32 for pure Mg and Mg-4.0 Zn-0.2 Ca alloy, respectively. In particular, the pH value is elevated to 11 from 7.4 for Mg-4.0Zn-2.0Ca alloy. This phenomenon can be explained as follows. The standard potential of coarse second phases of Mg-4.0Zn-1.5Ca and Mg-4.0Zn-2.0Ca alloys is higher than that of the pure Mg. Therefore, the selective attack occurred between α-Mg and the second phase, and the reaction in Hank's solutions is acute. Thus, the pH values of Mg-4.0Zn-1.5Ca and Mg-4.0Zn-2.0Ca alloys are rapidly increased. However, the uniform microstructure and lower reaction rates of pure Mg, Mg-4.0Zn-0.2Ca and Mg-4.0Zn-0.5Ca alloys cause a slow increase of pH values. After 48hrs immersion, bubbles are quite decreased which corresponds a slow reaction rate and leads a slow increase of the pH values for the samples. In addition, the increasing corrosion films including HA and other phosphates formed by the reaction during the immersion test can further reduce the reaction rates or degradation of the alloys [29].

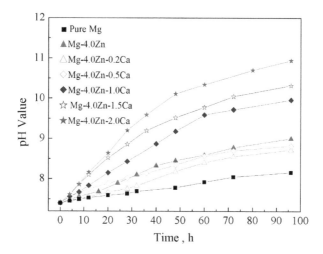

Fig. 10. pH variation of Hank's solution versus the immersion testing time for as-cast Mg-4.0Zn-xCa alloys.

According to the above-mentioned search, we found that Mg-4.0Zn-0.2Ca alloy have an excellent corrosion resistance among the Mg-Zn and the Mg-Zn-Ca alloys. Thus, the immersion test was only performance on the Mg-4.0Zn-0.2Ca alloy. The degradation rates of the alloy after 30-day immersion were listed in Table 5. The degradation rates of Mg-4.0Zn-0.2Ca alloy degraded were slower than pure Mg, which was adherent to the electrochemical results.

Mg and the Mg-4.0Zn-0.2Ca alloy degraded quickly, during the early stage of immersion in SBF, accompanied by the rapid formation of an insoluble protective corrosion layer, which retarded degradation. The degradation process of Mg-4.0Zn-0.2Ca alloy could be roughly summarized as follows: just after immersion in SBF solution, magnesium alloy react with fluids on the surface and get dissolved in the surrounding fluids. With the increasing time of immersion, more Mg^{2+}, Zn^{2+} and Ca^{2+} ions were dissolved into the solution, the local pH near the surface of the Mg could be >10[30]. As a result, a magnesium-containing calcium phosphate would precipitate from the SBF solution and deposited on the surface of the magnesium samples, per the following equation:

Anodic reaction:

$$Mg \rightarrow Mg^{+2} + 2e$$

Cathodic reaction:

$$2H_2O + 2e \rightarrow H_2 + 2OH^-$$

$$Mg^{+2} + 2OH^- \rightarrow Mg(OH)_2$$

$$PO_4^{3-} + Ca^{2+} + Mg^{2+} \rightarrow Mg_xCa_y(PO_4)$$

Moreover, when Mg2+, Zn2+ and Ca2+ ions were dissolved into the solution, phosphate-containing Mg/Ca insoluble protective layer was formed and tightly attached to the matrix. Previous studies [31] have shown that this corrosion layer promotes the osteo-inductivity and osteo-conductivity, predicting good biocompatibility of magnesium and retarded degradation. Therefore, it is proposed that the Mg2+, Zn2+ and Ca2+ released during degradation are safe. Hence, we come to the conclusion that the degradation of the Mg-4.0Zn-0.2Ca alloy was harmless and has good biocompatibility.

	E(V)	Current(mA/cm²)	V(mm/year)
As-cast	-1.60	2.67	2.05
Extruded	-1.57	2.43	1.98

Table 5. Corrosion potential, corrosion current and corrosion rate of Mg-4.0Zn-0.2Ca alloys

3.3.3 Corrosion morphology and products

The samples after electrochemical measurements were observed by SEM. The typical Surface morphology of Mg-Zn-Ca alloys after electrochemical measurements was shown in Fig.11. Corrosion attack on a large area was observed. At the same time, the filiform corrosion and pitting corrosion were found on the Mg-Zn-Ca alloys sample's surface after electrochemical measurements. The former mainly distributed on the grain boundary, and the latter mostly occurred in second phase location.

XRD patterns of the corrosion products on the surface of Mg-Zn-Ca alloys immersed in Hank's solution were presented in Fig.12. The XRD results suggest that magnesium hydroxide [Mg (OH) 2], other phosphates and hydroxyapatite (HA) were precipitated on the Mg-Zn-Ca alloys surface.

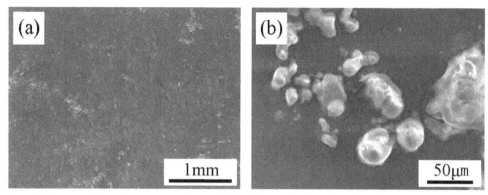

Fig. 11. The typical surface morphology of Mg-4.0 Zn-0.2 Ca alloy after electrochemical measurements: (a) macrostructure (b) microstructure

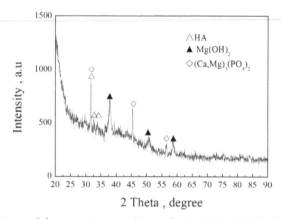

Fig. 12. XRD patterns of the corrosion products of Mg-4.0 Zn-0.2 Ca alloy immersed in Hank's solution.

3.4 Cytotoxicity assessments

The pH values of the extraction medium were measured, and only the values of Mg-1.0 Zn, Mg-2.0 Zn, Mg-3.0 Zn, Mg-4.0 Zn, Mg-4.0 Zn-0.2 Ca and Mg-4.0 Zn-0.5 Ca alloys were below than 8.0. That's mean all of these alloys have a potential probability used as the biomaterials. As the economic reason, only Mg-4.0 Zn-0.2 Ca was selected to evaluate the cytotoxicity through examining both the viability and morphology of L-929 cells in this study.

The morphologies of L-929 cells cultured in different extracts after 7 day incubation were shown in Fig.13. It could be seen that the cell morphologies in different extracts were normal and healthy, which was similar to that of the negative control. Fig.14 shows the RGR of L-929 cells after 2, 4 and 7 days of incubation. There was no significant difference between the RGR of cells in the extracts and those in the negative control. According to standard ISO 10993-5: 1999 [32], the cytotoxicity of these extracts was Grade 0-1. In other words, the Mg-4.0Zn-0.2Ca alloy has a level of biosafety suitable for in cellular applications.

Fig. 13. Morphologies of L-929 cells cultured for 7 day in different extraction media:(a) Negative control, (b) as-cast, (c) extruded.

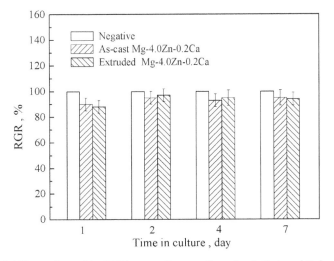

Fig. 14. Cell viability cultured in 100% extraction medium for 1, 2, 4 and 7 days.

The MTT tests were widely used cytotoxicity tests because they are easy, fast and cheap. But, in the case of Mg materials, the use of these test kits leads to false positive or false negative results. It is conceivable that Mg in the highly alkaline environment may be able to open the ring form of the tetrazolium salt and bind to it, which could lead to a change in colors similar to the formation of formazan in the case of the MTT tests with cells [20]. Thus the results of cytotoxicity tests conducted by MTT test will be higher than the true story. It is well known that the neutral red assay and the MTT assay exhibited almost identical reaction patterns for most test materials in L929 cells [33]. Thus, in this study, the neutral red kits were used to assay the cytotoxicity. Its measurement principle is based on the uptake of the vital dye neutral red into lysosomes of viable cells. Neutral red is accumulated because of the low intravesicular pH value. Lysosomes are, however, only one type of subcellular compartments which are acidified by ATP-driven proton pumps (V-type ATPase) and related low intravesicular pH value across vesicle membranes [34]. Accumulation of neutral red in acidic intracellular vesicles needs both ATP as a universal metabolic energy source for proton translocation against an electrochemical H^+-gradient and tightly sealed vesicle membrane to maintain potential differences. In an alkaline environment with Mg^{2+}, neutral red could lead to a change in color to yellow. But, the uptake of the vital dye neutral red into lysosomes was red color. The preliminary results of this test show that it seems not to be influenced by corroding Mg. Therefore, the neutral red assay may be regarded as a valid alternative method to determine cell viability, as it shows no interference with the corroding materials. The in-vitro cytotoxicity of Mg-4.0Zn-0.2Ca alloy was found to be Grade 0-1, indicating that the alloy was bio-safe.

3.5 In-vivo degradation

Furthermore, in order to further study the biocompatibility of Mg-4.0Zn-0.2Ca alloy, the in-vivo test was conducted on this new type magnesium alloy. Fig.15 showed the optical images of the cross-section of bone and magnesium implants after 3 months implantation. It could be seen that all the shapes of the magnesium implant had been changed from rod shape to irregular shape, indicating the implant was corroded by the body fluid, or the implant degraded in the body fluid. Meanwhile, a degradation layer or a reaction layer could be clearly found on the surface of the alloy implant, as indicated by D in Fig.15. In addition, newly formed bone was observed between the degradation layer and bone tissue around the magnesium alloy implants, as shown by N in Fig.15. The degradation rate was calculated according to the ratios of the cross section area of the residual implant to the original implant. After 3 months implantation, about 35-38% Mg-4.0Zn-0.2Ca alloy implant was degraded. Significant difference ($p < 0.05$) in the in-vivo degradation rates was observed.

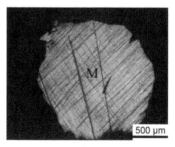

Fig. 15. Optical images of the cross-sections of Mg-Zn-Ca implants and bones after 3 months post implantation (M, metal; D, degradation layer; N, new bone; B,bone).

Fig.16 showed a high magnification microstructure of the bone implant interface after 3 months implantation by SEM. It could be clearly seen that the degradation layer was not dense, and many cracks were found. In order to reveal the chemical composition of the degradation layer, EDS was used to analyze the chemical composition of interface. The results were shown in Fig.16 (b). From the analysis results, it could be figured out that the degradation layer was mainly composed of carbon, oxygen, magnesium, calcium and phosphorous. However, the chemical composition was not homogeneous through the whole layer. At the position close to the Mg implant side, higher calcium content and higher Ca/P ratio were found. At the position close to the bone side, the calcium content was still high, but the Ca/P ratio became much smaller compared with at the position close to the Mg implant side. However, there was a sharp change in Mg content at the interface.

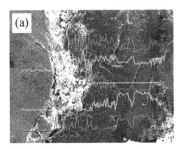 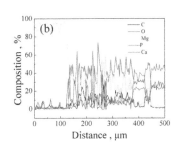

Fig. 16. (a) SEM microstructure of the interface between magnesium implant and bone interface after3 month post implantation, (b) EDS analysis patterns of implant and bone interface after 3 months post implantation.

3.6 Histological analysis

Fig.17 showed the tissue response to the Mg-4.0Zn-0.2Ca alloy pins implantation at 1, 2 and 3 months. It could be clearly seen that some lymphocytes were identified in histological tissue in 1 month after operation, but there was no visible evidence of multinucleated giant cells. After 2 months implantation, there was an active bone formation, which was evident by large number of new disorganized trabeculae. After 3 months implantation, new bone tissue was formed around the magnesium implant. In comparison with the histological microstructure obtained at the cortical bone near the implantation site, as shown in Fig.17, no difference could be found in the histological microstructure between the new bone and the cortical bone.

Magnesium alloys have attracted much attention as potential biodegradable bone implant materials due to their biodegradability in the bioenvironmental as well as their excellent mechanical properties such as high strength and an elastic modulus close to that of bone. In this paper, the in-vitro cytotoxicity and in-vivo biocompatibility of new kind of Mg-4.0Zn-0.2Ca alloy was studied. The cytotoxicity test indicated that the Mg-4.0Zn-0.2Ca alloy had no cytotoxicity. Rabbit implantation indicated that the Mg-4.0Zn-0.2Ca alloy did not cause any inflammation reaction. One month after operation, all magnesium implants were fixed tightly. There was no gap between the bone and the residual implant. Optical images from Fig.15 and SEM microstructure from Fig.16 showed clearly that there was a degradation layer formed on the surface of the magnesium implants. Histological images showed that new bone tissue was in contact with the magnesium implant through this degradation layer.

Line scanning in Fig.17 proved that large amount of Ca and P was found around the magnesium implant. In the present study, the histological analyses revealed that this magnesium-containing calcium phosphate degradation layer could promote or accelerate the new bone formation.

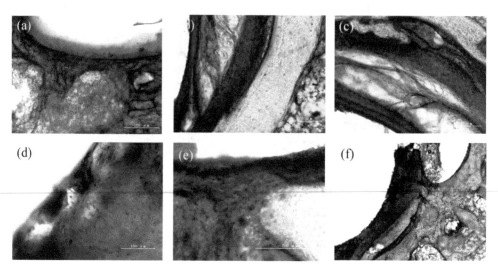

Fig. 17. Tissue response of the Mg-Zn-Ca alloy implantation at 1, 2 and 3 months (a) Mg-4.0Zn-0.2Ca 1 month, (b) Mg-4.0Zn-0.2Ca 2 month, (c) Mg-4.0Zn-0.2Ca 3 month, (d) cortical bone.

In-vivo degradation was a very complex process, so it was difficult to accurately assess the degradation rate of an implant material. F. Witte et al [35] found that the corrosion of a magnesium rod in the medullary cavity was not homogeneous in all cross-sections by used micro-computed tomography. In our study, the cross-section area of the residual implant was calculated to describe the degradation rate of the Mg implant. Due to the inhomogeneous corrosion of a magnesium rod in the medullary cavity, the calculated degradation rate based on the images was similar to the real in-vivo degradation rate of magnesium implants.

In-vivo degradation, compare with other metal implants, is an ultimate merit of magnesium alloy. After implantation in the rabbit, Mg-4.0Zn-0.2Ca alloy would be reacted with body fluid on the surface and get dissolved in the surrounding body fluid. At first, the released Mg^{2+}, Zn^{2+} and Ca^{2+} could be absorbed by the surrounding tissues and excreted through the gastrointestinal route and the kidney. However, with the increasing time of implantation, more Mg^{2+}, Zn^{2+} and Ca^{2+} ions are dissolved into the solution, the local pH near the surface of the Mg implants could be >10[36]. As a result, an insoluble magnesium-containing calcium phosphate would be precipitated from the body fluid on the surface of the magnesium implant and tightly attached to the matrix, which retarded degradation. In addition, the corrosion layer on the Mg-4.0Zn-0.2Ca alloy contained Mg, Ca and P, which could promote osteoinductivity and osteoconductivity, predicting good biocompatibility of magnesium. Therefore, it is proposed that the Mg^{2+}, Zn^{2+} and Ca^{2+} released during degradation should be safe.

4. Conclusion

In this paper, we developed ternary Mg-Zn-Ca alloys as biodegradable materials. The following conclusions can be drawn.

The mechanical properties of the as-cast Mg-Zn-Ca alloys can be tailored by the Zn and Ca content. The tensile strength can be increased form 105MPa to 225Mpa, and the elongation can be increased from 4.2% to17 %.

The in-vitro degradation of Mg- Zn- Ca alloys revealed that Zn and Ca not only elevated the corrosion potential of the magnesium alloys, but also influence their corrosion current. A protective layer of $Mg(OH)_2$ and other Mg/Ca phosphates was formed on the surface of Mg- Zn- Ca alloys when immersed in SBF solution ,which declined the degradation rate.

In vitro cytotoxicity assessments indicated that Mg-4.0 Zn-0.2Ca alloy did not induce toxicity in L-929 cells and are suitable for biomedical applications.

Implanted in rabbits, the alloy did not induce inflammation reactions or affect the new bone formation. We could draw a conclusion that the Mg-4.0Zn-0.2Ca alloy had good biocompatibility.

5. References

[1] M.M.Avedesian, H.Baker. editors. ASM Specialty Handbook, Magnesium and Magnesium Alloys, ASM International Materials Park,USA, Ohio, 1999,p. 14.

[2] F. Witte, V. Kaese, H. Haferkamp, E. Switzer, A. Meyer-Lindenberg, C.J. Wirth. Biomaterials 26 (2005) 3557.

[3] P.S. Mark, M.k. Alexis, H. Jerawala, D. George. Biomaterials.27 (2006) 1728.

[4] L.P. Xu, G.N. Yu, E.L. Zhang, F. Pan, K.Yang. J. Biome. Maters. Res .83A(2007)703.

[5] R.Zeng, W. Dietzel, F. Witte,N. Hort, C. Blawert. Adv. Biomate.10 (2008)B3.

[6] X.N. Gu , Y.F. Zheng, Y.Cheng, S.P. Zhong, T.F. Xi. Biomaterials.30 (2009) 484..

[7] Z.J. Li, X.N. Gu, S.Q. Lou, Y.F. Zheng. Biomaterials.29 (2008)1329.

[8] Y.Wan, G. Xiong,H.Luo,F. He,Y. Huang,X. Zhou. Mater. Design 29 (2008)2034.

[9] H.Tapiero, K.D.Tew. Biomed Pharmacother. 57 (2003) 399.

[10] S.X. Zhang, X.N. Zhang, C.L. Zhao, J.N. Li, Y. Song, C.Y. Xie. Acta Biomate.6(2010)626.

[11] P.M. Jardima , G. Solórzano, J.B. Vander Sande. Mater Scie Eng A .381 (2004) 196.

[12] G. Ben-Hamu , D. Eliezer, K.S. Shin. Mater Sci EngA.447 (2007) 35.

[13] H.Tapiero, K.D.Tew. Biomed Pharmacother. 57 (2003) 399.

[14] Y .Ortega, M.A. Monge, R.Pareja. J Alloys Compd. 463 (2008) 62

[15] E. Zhang, D.S.Yin, L.P.Xu, L.Yang, K.Yang. Mater.Sci.Eng.C. 29(2009) 987.

[16] L.P. Xu, G.N. Yu, E. Zhang,F. Pan, K.Yang. J. Biomed. Mater. Res. A. 83A(2007)703.

[17] H.X.Wang, S.K.Guan, X .Wang, C.X.Ren,L.G.Wang. Acta Biomate.6(2010)1743.

[18] L.Mao,Y.Wang,Y.Wan, F. He, Y.Huang. Heat Treat. Metal. 34(2009)19.

[19] X.N.Gu, Y.F. Zheng, S.P. Zhong, T.F. Xi, J.Q Wang, W.H. Wang. Biomaterials 31 (2010) 1093.

[20] J.Fischer, M.H.Prosenc, M.Wolff, N.Hort, R.Willumeit, F.Feyerabend.Acta Biomate. 6(2010) 1813.

[21] American Society for Testing and Materials. ASTM-G31-72: standard practice for laboratory immersion corrosion testing of metals. In: Annual Book of ASTM Standards. Philadelphia, PA: American Society for Testing and Materials; 2004.

[22] A. Datta, U.V. Waghmare, U. Ramamurty, Acta Mater. 56 (2008) 2531.

[23] H. Okamoto. J. Phase Equilibria. Diffus.15 (1994) 129.

[24] X.Gao, S.M.Zhu, B.C.Muddle, J.F.Nie, Scripta Mater. 53 (2005) 1326.

[25] J.C. Oh, T. Ohkubo, T. Mukai, K. Hono, Scripta Mater. 53 (2005) 675.

[26] L.Geng, B.P. Zhang, A.B. Li, C.C. Dong, Mater. Lett. 63 (2009) 557.

[27] Y.C. Xin , C.L.Liu , K.F.Huo, G.Y. Tang, X.B.Tian, P.K.Chu. Surf. Coat.Tech. 203 (2009) 2554.

[28] Y. C. Xin, K. F. Huo, T. Hu, G. Y. Tang, P. K. Chu. Acta Biomate. 4 (2008) 2008.

[29] Y. C. Xin, K. F. Huo, T. Hu, G. Y. Tang, P. K. Chu. J. Mater. Res. 24 (2009) 2711.

[30] Y. C. Xin, T. Hu, P. K. Chu. J. Electrochem.Soc.157(2010)C238.

[31] F. Witte, F.Feyerabend, P.Maier, J.Fischer, M. Störmer, C.Blawert, et al. Biomaterials. 28(2007)2163.

[32] ANSI/AAMI. ISO 10993-5: 1999. Biological evaluation of medical devices. Part 5. Tests for cytotoxicity: in vitro methods. Arlington, VA: ANSI/AAMI.

[33] H. Schweikl, G .Schmalz. Eur. J. Oral. Sci.104(1996) 292.

[34] N. Nelson. J Exp Biol 172(1992) 19.

[35] F.Witte, J. Fischer, J.Nellesen, H.A. Crostack, V. Kaese, A. Pisch, F. Beckmann, H. Windhagen. Biomaterials. 27(2006)1013.

[36] A. Simaranov, I. Sokolova, A. Marshakov, Y. Mikhailovskii. Prot. Metal. 27(3) (1991) 329.

9

Decellularization, Stabilization and Functionalization of Collagenous Tissues Used as Cardiovascular Biomaterials

Birzabith Mendoza-Novelo[1] and Juan Valerio Cauich-Rodríguez[2]
[1]*Universidad Politécnica de Juventino Rosas*
[2]*Centro de Investigación Científica de Yucatán*
México

1. Introduction

Cardiovascular diseases are a worldwide problem being a significant cause of morbity and mortality every year. Patients requiring heart valve replacements include those exhibiting degenerative valvular diseases and rheumatic fever. The pathological processes include stenosis, fibrosis, myxoid change and calcification. The fibrosis causes a reaction to normal haemodynamic while the myxoid change reduces tensile strength of the valve due to replacement of dense collagenous tissue by loose tissue rich in glycosaminoglycans. Moreover, these pathologies can be observed in normal valves or fibrotic valves (Lindop, 2007).

Fortunately, the development of cardiovascular prostheses, either synthetic or biological, has allowed to increase life expectancy and has improved the quality of life of patient requiring either heart valves (Flanagan & Pandit, 2003; Schoen & Levi, 1999; Vesely, 2005) or vascular grafts (Matsagas et al., 2006; Monn & West, 2008; Schmidt & Baier, 2000). The implant technology for cardiovascular systems made use of raw materials of different origins. For example, metallic materials and synthetic polymers have been widely used in mechanical valves for the replacement of diseased heart valves. However, some complications such as alterations in the hemodynamical function and thrombus formation have been found (Zilla et al., 2008).

Biological prostheses provide some answers to these complications, although the bioprostheses do not fulfil their objectives satisfactorily, since they display others complications once implanted. The complications of tissue valves include calcification, remnant tissue immunogenicity, inflammatory degradation, mechanical damage and lack of repair (Zilla et al., 2008). Therefore, the need for safe, economic, physiologically acceptable and viable biomaterial has motivated the modification of collagen-rich tissues.

Collagenous tissues are alternative raw materials for the manufacture of medical devices due to their physical and biomechanical properties. These tissues promote cell interactions, exhibit good ion and macromolecular binding capacity in addition to their electrostatic, hemostatical and immunological properties (Li, 2007). Since 1960s, perichardial tissues and the porcine heart valves are two of the most widely used biological tissues in the construction of cardiovascular devices. The introduction of these biological biomaterials was linked to the tissue crosslinking to increase durability. However, due to some complications in the stabilized tissue, several post-crosslinking protocols have been proposed to address these complications. More recently, biological scaffolds derived from acellular tissue has been used in tissue engineering and regenerative medicine.

Therefore, this chapter deals with the processing of collagenous tissue for the preparation of cardiovascular biomaterials. The processing techniques include the extraction of cellular and nuclear material by various decellularization methods, the preservation of tissue through of crosslinking reactions, hydrogen-bond interactions or interstitial space filling, and the functionalization or the blocking of free groups with various low molecular and macromolecular substances.

2. The biomaterial choice

The replacement of damaged organs or tissues is one of the objectives of the biomaterials science. For this, natural or synthetic materials can be used for example in the cardiovascular field in the manufacture of heart valves and vascular grafts. The success of a device depends not only on the type of biomaterials but also on a set of acceptable characteristics such as biocompatibility, biostability, haemocompatibility, anti-trombogenecity, resistance to degradation and calcification.

Among those biomaterials that can fulfil these requirements, natural tissues are good candidates and that is why they have been under investigation in the past fifty years.

2.1 Composition and sources of natural tissue

Natural tissue biomaterial can be obtained from either animal-derived tissue (xenograft) or human-derived tissue (homograft). However, due to the limited availability of autografts, animal-derived tissues are, in many cases, the first choice for cardiovascular biomaterials. Animal derived tissues widely used as biological biomaterials include perichardial tissue from various sources such as cows, calves and ostrich in addition to pig aortas.

Tissue-derived biomaterials are mainly comprised of collagen in addition to the tissue extracellular matrix (ECM) which is a complex mixture of structural and functional proteins such as collagen, proteoglycans, glycoproteins, elastin, metalloproteins, etc. Collagen, being the main structural protein, is a polypeptide that contain amino ($-NH_2$), carboxylic acid ($-COOH$) and hydroxyl ($-OH$) functional groups as substituents, and together with the amide bonds in the polymer backbone form the reactive centers. The repetitive unit in the polymer backbone of collagen and the amino acid residues as side group are depicted in figure 1.

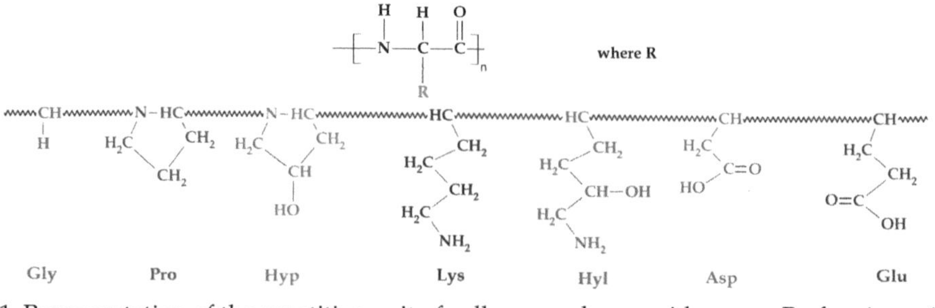

Fig. 1. Representation of the repetitive unit of collagen and some side group R of amino acid residue

The crystal lattice of collagen fibers are embedded in an amorphous matrix. The amorphous matrix is composed mainly by glycosaminoglycans as proteoglycans (sulphated glycosaminoglycans bound to proteins). In this matrix, in addition to the fibers, tissue cells and interstitial fluid (water or electrolytes) are embedded. The glycosaminoglycans are negatively charged polysaccharides of varying degrees of complexity. The glycosaminoglycan polymers consist of repeating disaccharide units, usually consisting of a hexosamine and an uronic acid (Yeung et al., 2001). The charged negatively units contribute to the elasticity and hydration of the tissues (Mavrilas et al., 2005), but may attract counter-ions, which could intervene in the processes of calcification of bioprostheses. The repetitive disaccharide unit of glycosaminoglycans mainly presents in native bovine perichardium is shown in figure 2.

Hyaluronan

Chondroitin sulphate

Dermatan sulphate

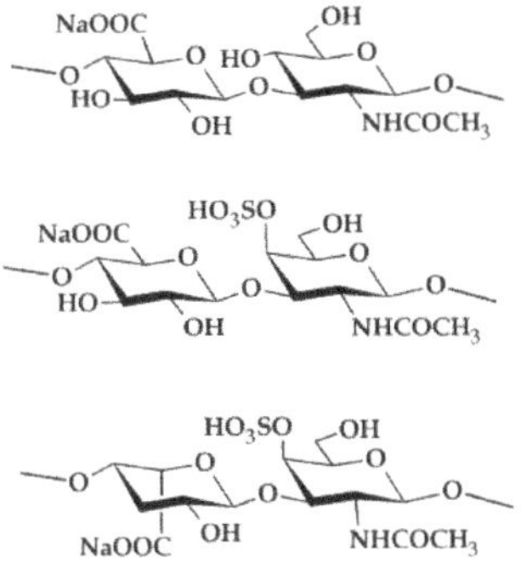

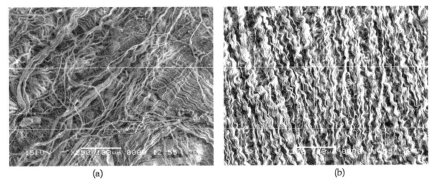

Fig. 2. Repetitive disaccharide units of common glycosaminoglycans in bovine perichardial tissue

The different soft tissues including cartilage, tendons, ligaments, skin and perichardium have the capacity of support mechanical load of variable magnitude. Therefore, the properties of the tissue depend on the number and the arrangement of collagen fibers, which can be parallel or perpendicular to the surface or randomly distributed in the matrix. The hierarchical nature of collagen confers to the tissue its structural complexity. The fibrous nature of bovine perichardial tissue is revealed in figure 3. In perichardial tissue, a multi-laminate structure is observed with difference in both serosa (Fig. 3b) and rugosa surface (Fig. 3a).

Fig. 3. SEM micrographs for the fibrosa (a) and the serosa surface (b) of native bovine perichardium

2.2 Properties of collagenous tissues

Collagen-rich tissues are composed of 75% of collagen, 20% of mucopolysaccharides and water, although elastin can be found in less than 5% (Cauich-Rodríguez, 2008). All these tissue components maintain the structural and functional integrity of the composite tissue. Some mechanical properties of collagenous tissues are shown in table 1.

Tissue	Property	Value	Reference
Bovine perichardium	Tensile strength	10.9 MPa	Lee at el., 1989
	Strain at rupture	33.0 %	
	Tissue modulus	58.2 MPa	
Canine perichardium	Tensile strength	18.4 MPa	Lee & Boughner, 1981; Wiegner & Bing, 1981
	Strain at rupture	21.4 %	
	Tissue modulus	198 MPa	
Human perichardium	Tensile strength	2.51 MPa	Lee & Boughner, 1985
	Strain at rupture	34.9 %	
	Tissue modulus	20.4 MPa	
Porcine aortic valve leaflet	Tensile strength	6.25 MPa	Lee at el., 1984
	Strain at rupture	30.8 %	
	Tissue modulus	54.6 MPa	

Table 1. Mechanical properties of some collagenous tissue

The thermal transitions experienced by materials with amorphous and/or crystalline regions are also observed in the collagenous tissue. When the biomaterial is heated, its specific volume increases, exhibiting the glass transition of amorphous regions and the fusion of crystalline collagen fibers to a temperature higher than the glass transition temperature (Li, 2007). The melting temperature of collagen fibers is an irreversible process and is often referred in the literature as the denaturation temperature (T_d) or shrinkage temperature (T_s). In fact, the denaturation temperature is widely used as an indicator of the tissue stabilization.

The collagenous tissues require chemical or physical treatments in order to be preserved or stabilized. In fact, the introduction of cardiovascular bioprostheses in 1960s was linked to the chemical fixation of porcine aortic valves or bovine perichardial tissue with glutaraldehyde. This process produces a non-living material without the capability of intrinsic repair as native tissue does after some structural injury (Flanagan & Pandit, 2003). The processed tissue tends to fail in modes related to the remnant immunogenicity, inflammatory degradation, mechanical damage and pannus overgrowth (Zilla et al., 2008). In general, the stabilization of collagen-rich tissue is achieved by direct binding of functional groups to amino acid residues from collagen by coupling agents or by the linkage between the functional groups on collagen and various chemical agents. Both processes are referred in literature as the fixation or crosslinking processes. While the crosslinking agents make durable, stable and resistant tissues, the crosslinking density and the chemical process seems to have an effect on some of the major disadvantages of bioprostheses, such as calcification (Zilla et al., 2008). For this reason, a large number of crosslinking agents have been suggested with the aim of obtaining bioprosthesis that fulfill successfully its function. In addition to this treatment, there are reports on the post-crosslinking and pre-crosslinking treatments in order to reduce the calcification of biomaterial and in order to prepare porous biomaterials as scaffolds for tissue engineering.

3. Decellularization of tissues

The concept of decellularization is referred as the extraction of cellular components from natural tissues of human or animal origin. Different approaches have been reported as effective procedures to remove cells from xenogeneic and allogeneic collagenous tissue with the aim of removing cellular antigens and procalcifying remnants while the extracellular matrix (ECM) integrity is preserved as much as possible (Schmidt & Baier, 2000). The combination of chemical, physical and enzymatic methods destroys the cell membrane and removes nuclear and cellular material (Gilbert et al., 2006). The remaining acellular ECM will be a complex mixture of structural and functional proteins, glycoproteins and glycosaminoglycans arranged in a three-dimensional architecture. However, some mechanical and structural alterations on the ECM can be induced during the decellularization process.

3.1 Effect of decellularization treatment on tissue properties

A biomaterial or scaffold for tissue engineering should provide not only mechanical support for the cell proliferation but also they must be versatile to give the required anatomical shape (Kidane et al., 2009). The decellularization of collagenous tissues has been explored as the ECM may serve as appropriate biological scaffold for cell attachment and proliferation. However, alterations both in the structural composition and in the mechanical properties of the remaining ECM can be induced during the decellularization protocols. The mechanical integrity can be affected and it may be associated either to the denaturation of the collagen triple helix or to the loss of macromolecular substances such as glycoproteins.

The efficiency of a given decellularization method and their effects on the properties of animal tissues must be studied in a specific manner due to compositional and structural differences (Gilbert et al., 2006). For example, the decellularization of porcine heart valve with sodium dodecyl sulphate, an anionic detergent, appeared to maintain the critical mechanical and structural properties of the valves leaflets (Liao et al., 2008) while decellularization of bovine perichardium with sodium dodecyl sulphate caused irreversible swelling, resulted in a reduction of the denaturation temperature (Courtman et al., 1994; García-Páez et al, 2000) and caused a reduction of almost 50% on tensile strength when compared to native tissue and tissue treated with Triton X100, a non-ionic detergent (Mendoza-Novelo & Cauich-Rodríguez, 2009).

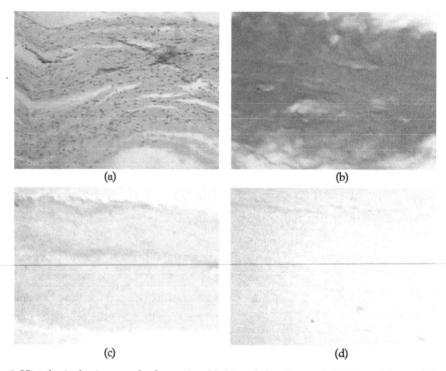

Fig. 4. Histological micrographs for native (a),(c) and decellurized (b),(d) perichardial tissue in H&E (top) and alcian blue (bottom) staining

It has been proposed that an anionic detergent binds to proteins, increases negative charges and results in tissue irreversible swelling (Courtman et al., 1994). In addition, a highly negative charged perichardial tissue has been associated to a higher tendency to tissue calcification (Jorge-Herrero et al., 2010). Due to these adverse effects, non-ionic detergents are preferred over ionic surfactants in the decellularization process of perichardial tissue. However, there are some issues related to the use of aromatic (phenolic) or non-aromatic (non-phenolic) non-ionic detergents used in the decellularization process. For example, the biodegradation products of derivatives of non-ionic detergents such as alkylphenol ethoxylates have been associated to toxicity (Argese et al., 1994) and estrogenic effects (Soto et al., 1991; Jobling & Sumpter et al., 1993). Figure 4 shows the histological results for bovine perichardial tissue decellularized with a non-aromatic non-ionic detergents. In this case, a reduction in the cell nuclei present in bovine perichardial tissue and a decrease in the glycosaminoglycan content after decellularization treatment were observed (Mendoza-Novelo et al., 2011).

In addition to tissue decellularization with nonionic surfactants, reversible swelling has also been studied. In this case, the reversible alkaline swelling did not change the three-dimensional architecture of native bovine perichardium. This means that the laminar structure and fibrous nature of the native perichardial tissue were maintained after decellularization although the opening of the interfibrilar spaces was observed. The reversible alkaline swelling cause a reversible change in the tissue thickness i.e. increased 45% after swelling step, but the tissue original thickness was regained after deswelling step.

However, the alkaline treatment altered the perichardial tissue stress relaxation behaviour (Mendoza-Novelo et al., 2011).

3.2 Pre-treatments (pre-crosslinking) methods to reduce tissue calcification

Bovine pericardium undergoes several treatments prior to crosslinking with the aim to improve its biocompatibility, to reduce immunogenicity, to decrease its tendency to calcification, to promote neo-vascularization and infiltration, and to increase cell adhesion and proliferation. Some of the pre-treatments proposed in the literature to reduce calcification of cardiovascular bioprostheses are showed in the table 2. It has been reported that with the treatment of bioprostheses with sodium dodecyl sulphate and Triton™X-100 most of the acidic phospholipids are extracted resulting in the initial suppression of calcification in the cell membrane (Schmidt & Baier, 2000).

Pre-treatment	Anti-calcification action mode	Reference
Surfactants	Removal of acidic phospholipids	Schmidt & Baier, 2000; Chang et al., 2004
Alcohols	Removal of phospholipids and cholesterol Alteration in the collagen conformation Cellular death Removal of Cardiolipin	Vyavahare et al. 1997; Pfau et al. 2000; Pathak et al., 2002

Table 2. Tissue pretreatment in order to reduce the bioprostheses calcification

The pretreatment of collagen-rich biomaterials with different concentrations of ethanol may prevent calcification through the extraction of phospholipids and cholesterol but causes a permanent alteration in the collagen conformation (Schmidt & Baier, 2000). Additionally, this treatment affected the interaction of the tissue with water and lipids and increased the resistance of the tissue to the action of collagenase. Several high molecular weight alcohols have been used in order to remove cellular components that contain elements responsible for the calcification (Pathak et al., 2002). The pretreatment with 50% ethanol for 5 min reduces fibrosis of bovine pericardium implanted in the aorta of sheep as a result of cell death and cardiolipin removal more than the phospholipids extraction (Vyavahare et al., 1997). Mixtures of chloroform/methanol have also been effective in reducing tissue calcification (Jorge-Herrero et al., 1994).

4. Stabilization of tissues

The stability of tissues is increased by physical or chemical crosslinking. The fixation enhances tissue stability, inhibits autolysis, allows a prolonged shelf-life, and allows a surgeon to have medical devices of various sizes readily available for implantation (Schoen & Levy, 1999). The chemical treatments also mitigate immunogenicity while maintaining both thromboresistance and antimicrobial sterility but greatly influence their degradation and calcification. However, tissue calcification is multifactorial phenomenum where chemical crosslinking is considered just one of these factors. In fact, the alteration in the electrical charge that exists in the perichardial tissue surface has been associated to the calcification (Jorge-Herrero et al., 2010). Several crosslinking techniques have been suggested as the ideal procedure to stabilize the collagen structure while maintaining their physical and natural shaping. The structure and name of some chemicals used as crosslinking agents for collagenous tissue are shown in table 3.

Structure of stabilization agents	Name	Reference
	Glutaraldehyde	Olde Damink et al., 1995; Duncan & Boughner, 1998; Langdon et al., 1999
	Ethyl-aminopropyl carbodiimide (EDAC) hydroxysuccinimide (NHS)	Lee et al., 1996; Everaerts et eal., 2004; Mendoza-Novelo & Cauich-Rodríguez, 2009
	Glycerol diglycidyl ether	Lee et al., 1994
	Ethylene glycol diglycidyl ether, n=1; Poly(ethylene oxide) diglycidyl ether, n=22	Tu et al., 1993; Sung et al., 1996; Zeeman et al., 2000
	Triglycidylamine	Conolly et al., 2005; Sack et al., 2007; Rapoport et al., 2007
	Hexamethylene diisocyanate	Naimark et al., 1995; Olde Damink et al., 1995; Nowatzki & Tirrel, 2004
	Disuccinimidyl suberate	Pathak et al., 2001
	Genipin	Sung et al. 1999; Sung et al., 2000
	Proanthocyanidin, Procyanidins	Han et al., 2003; Zhai et al., 2009
	Reuterin	Sung et al., 2002; Sung et al., 2003

	Tannic acid	Isenburg et al., 2004; Isenburg et al., 2006

Table 3. Some chemical agents used for the stabilization and fixation of biological tissues

4.1 Tissue crosslinking with glutaraldehyde

The procedure most studied and exploited in the manufacture of tissue valve includes the crosslinking with glutaraldehyde, which is also widely used as tanning agent in the leather industry. Glutaraldehyde is an important reagent in the biomedical field and has been used as crosslinking agent in the preparation of collagen-rich biomaterials or for the immobilization of enzymes or cell fixation.

Glutaraldehyde is an efficient agent for the crosslinking of collagen matrix because it react relatively quickly and because is able to join separate protein molecules by means of the amino groups abundantly present in collagen. Glutaraldehyde is a cheap and water soluble five-carbon bifunctional aldehyde that in aqueous solution consists of a mixture of free aldehyde, mono and dihydrated monomeric glutaraldehyde, monomeric and polymeric cyclic hemiacetals and various α, β unsaturated polymers (Whipple & Ruta, 1974). This means that glutaraldehyde itself forms a number of different reactive species and that these species may also react in different ways, rendering a highly crosslinked network. Glutaraldehyde crosslinking has been and is still applied to most of the experimental and clinical bioprostheses. This process consists in blocking the ε-amino groups of lysine in the protein through imino bond formation. The contribution of the glutaraldehyde as sterilization and crosslinking agent is partly due to its hydrophobicity and hydrophilicity, allowing it to penetrate both aqueous media and in the cell membrane. However, in the manufacture of bioprostheses, the use of glutaraldehyde has led to many disadvantages associated with the residual free aldehyde groups. Table 4 shows some of the problems associated with glutaraldehyde tissue crosslinking and some solutions that have been suggested to solve them.

In aqueous solution, the glutaraldehyde is presented as a mixture of free aldehyde, mono and dihydrate glutaraldehyde monomer, monomeric and polymeric cyclic hemiacetals, and several alpha or beta unsaturated polymers (Monsan et al., 1975). In turn, this heterogeneity of chemical species leads to a heterogeneous crosslinking. In addition, high concentration of glutaraldehyde promotes rapid surface crosslinking in the tissue (Olde-Damink et al., 1995), creating a barrier that impedes or prevents the diffusion of more glutaraldehyde within the biomaterial. In order to avoid this, the use of low concentrations has been suggested (Khor, 1997). It has also been proposed glutaraldehyde protection as a monomer by the formation of di-acetals, between glutaraldehyde and alcohols in acidic medium (Giossis et al., 1998).

The fixation reaction was carried out by the exposure of the tissue balanced with glutaraldehyde acetals solutions to triethylamine vapours. This process allowed the diffusion of the non-reactive glutaraldehyde into the tissue, minimized the formation of polymeric glutaraldehyde and reduced the waterproofing (hydrophobicity) at the tissue surface (Yoshioka & Giossis, 2008).

The conditions of the crosslinking reaction (pressure for instance) have been varied with the aim of improving the biomechanical properties of bovine perichardium. The crosslinking of bovine perichardium with glutaraldehyde at a pressure of 4 mm Hg (low pressure) both statically and dynamically (1.2 Hz) has been reported. By comparing the properties of crosslinked bovine perichardium, the dynamically crosslinked tissue showed a very similar extensibility to native biomaterial (non-crosslinked) in contrast to statically crosslinked tissue, which showed a higher extensibility, while no differences were reported in other mechanical properties (Duncan & Boughner, 1998). The bovine perichardial fixation with glutaraldehyde under biaxial static pressures (~225 and ~1875 mmHg) has been proposed. The bovine perichardium treated at high pressure showed an increase in stiffness and almost isotropic behaviour, while low pressure-treated bovine perichardium preserved the anisotropy exhibited by the native tissue (Langdon, et al., 1999). Porcine valves have also been subjected to crosslinking at high pressure (80 mm Hg), low and zero pressure. In this case, it was reported an increase in the rigidity of the leaflets fixed under low pressure and the preservation of geometric corrugations and undulations of the native tissue when the leaflet were fixed without pressure (Lee et al., 1984).

Heat treatment during glutaraldehyde fixation has also been reported. The thermal treatment at 50°C showed an anti-calcifying effect which was attributed to structural changes in collagen or lipid extraction by heat treatment (Carpentier et al., 2001).

4.1.1 Post-treatments after glutaraldehyde fixation

The residual unbounded aldehyde groups that remain in the tissue after glutaraldehyde fixation process have been associated with degenerative phenomenum on different bioprosthesis. The grafting of different molecules on collagenous tissues treated with glutaraldehyde has been an answer to these disadvantages.

The grafted molecules are incorporated in order to block free aldehyde groups and thus to reduce or to neutralize both cytotoxicity and calcification. Some surface modification procedures of crosslinked collagenous tissues are described in table 5.

It is known that nitric oxide releasing compounds can improve the biocompatibility of blood-contacting medical devices (Frost et al., 2005; Masters et al., 2005). Two common nitric oxide generating substances immobilized on synthetic polymers are diazeniumdiolates and S-nitrosothiols (Frost et al., 2005). In the same line of thought, surface modification of polymeric materials, such as PET or PU, with thiol compounds is interesting as it might exchange nitric oxide with endogenous donors such as S-nitrosothiols that already circulate in blood (Gappa-Fahlenkamp et al., 2004; Gappa-Fahlenkamp & Lewis, 2005).

The thiol groups on the polymer allowed the exchange reaction with S-nitroso serum albumin and then, the release of nitric oxide to inhibit platelet adhesion on the polymeric surfaces (Duan & Lewis, 2002). This approach has been proposed in perichardial tissue biomaterial by using L-cysteine as thiol compound (Mendoza-Novelo & Cauich-Rodríguez, 2009). One additional advantage of L-cysteine grafting on glutaraldehyde-crosslinked perichardial tissue is that free aldehyde groups will be diminished or even eliminated on the tissue allowing its detoxification. A schematic representation of grafting of collagenous tissue with L-cysteine is described in the figure 5. Similar approaches with other amino acids have been suggested in order to provide non-cytotoxic tissue biomaterial and biomaterial with reduced calcification, as it is shown in table 5. In this table, it is also included molecules of biological importance as well as peptide sequences used to improve cell adhesion after the fixation treatment.

Type of molecule		Effects on biomaterial	References
Macromolecules	Hyaluronic acid -	Reduce the calcification of glutaraldehyde-treated tissue	Ohri et al., 2004
	Heparin-	Reduce the calcium deposition and the cytotoxicity of glutaraldehyde-treated	Lee et al., 2000
	Poly(ethylene glycol)-	Inhibit the platelet surface attachment and spreading and decrease the calcification of glutaraldehyde-treated tissue	Vasudev & Chandy, 1999; Park et al., 1997
	RGD peptides-	Enhace the adhesion and proliferation of human mesemchymal stem cells on acellular tissue	Dong et al., 2009
Amino acids	L-arginine, L-glutamine, L-lysine, L-glutamic acid, L-cysteine -	Reduce the protein adsorption and platelet adhesion of glutaraldehyde treated tissue. However, BP treatment with amino acids does not effectively prevent calcification. Incorporation of thiol moieties to the tissue	Jorge-Herrero et al., 1996; Jee et al., 2003; Mendoza-Novelo & Cauich-Rodríguez, 2009
Acids	Homocysteic acid -	Reduce toxicity but does not affect the stability of glutaraldehyde-treated tissue	Stacchino et al., 1998
	Amino oleic acid -	Inhibit the calcification of glutaraldehyde-treated tissue	Chen et al., 1994

Table 5. Molecules grafted on crosslinked bovine perichardial tissue

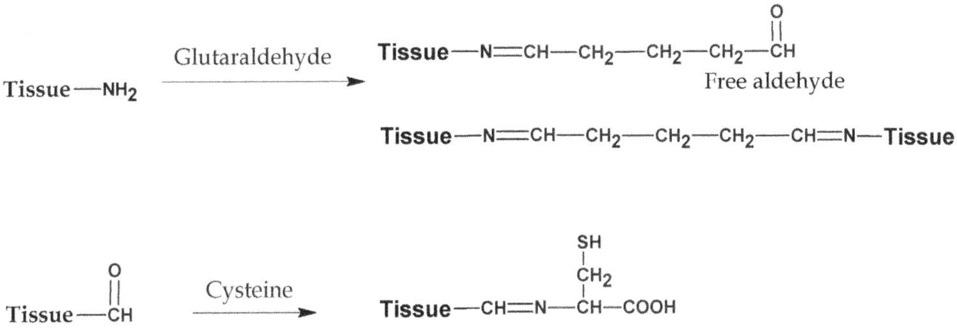

Fig. 5. Schematic representation of tissue crosslinking with glutaraldehyde and chemical coupling of L-cysteine

4.2 Tissue crosslinking after carboxylic group activation

Due to the problems associated with the use of glutaraldehyde, various non-aldehyde alternative methods have been developed to stabilize and post-treat tissues. The crosslinking agents used in collagen-rich biomaterials can use both primary amino groups and acid groups of polypeptide chains. Historically, a water soluble carbodiimide (1-ethyl-3-(3-dimethyl amino propyl) carbodiimide / EDAC) was first used for the modification of carboxylic groups in proteins for peptide synthesis (Sheehan & Hlavka, 1956) and to promote crosslinking in gelatin (Sheehan & Hlavka, 1957).

The mechanism for the reaction between carboxylic groups and EDAC leading to amide bond formation is as follows: The addition of a carboxylic acid diimide produces an isourea ester, an O-acyl isourea. The intermediate O-acyl isourea is an activated carboxylic acid derivative with similar reactivity to an anhydride or acyl halide, and can be subjected to a subsequent nucleophilic substitution by an amine yielding a dialkyl amide and urea (Carraway & Khosland, 1972). Because carbodiimide is just a coupling agent, when used to crosslink collagen in the absence of agents with dual functionality, only promotes the formation of an amide bond between carboxylic acid and amino reactive groups present in the tissue, as depicted in figure 6.

Fig. 6. Schematic representation of tissue crosslinking with EDAC and NHS

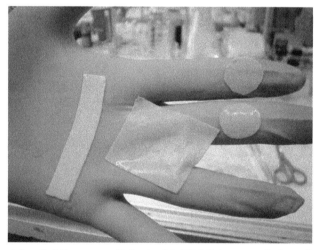

Fig. 7. Bovine perichardial tissue crosslinked with EDAC (rectangules) or glutaraldehyde (circles)

This requires that the activated carboxylic groups be close enough to the amino groups to achieve direct bonding (amide bond formation). The carbodiimides are hydrolyzed rapidly in aqueous solution and the intermediate O-acyl isourea is extremely unstable producing a low crosslinking.

The crosslinking density and the shrinkage temperature of bovine perichardium treated with EDAC had values lower that a control of bovine perichardium fixed with glutaraldehyde (Mendoza-Novelo & Cauich-Rodríguez, 2009). However, the use of the N-hydroxysuccinimide (NHS) during crosslinking with EDAC improved the stabilization of tissue due to the formation of a stable intermediate compound after reaction of the NHS with carboxylic groups or isourea O-acyl intermediate (Lee et al., 1996). Such is the case reported for porcine aortic valves crosslinked by a two-step method. These steps included the blocking of the free primary amino groups of collagen with butanal and the crosslinking with Jeffamines™ of different molecular weights by activating the carboxylic acid groups with EDAC and NHS. This process led to a decrease in calcification (subcutaneous implantation in rats) of engineered tissue (Everaerts et al., 2004).

The appearance of bovine perichardial tissue crosslinked with glutaraldehyde and EDAC is shown in figure 7.

4.3 Tissue crosslinking with epoxy compounds

The chemistry of epoxy groups, cyclic ethers of three members, has also been explored and applied in the fixation of tissue. Polyepoxide compounds or epoxy bifunctional polyether react with amino groups from collagen opening the terminal epoxide ring (Tu et al., 1993; Lee at al., 1994; Khorn, 1997). This reaction is nucleophilic and can be carried out under acidic conditions (highly reactive protonated epoxy) or alkaline (amine at its most nucleophilic). In this case, the modification of swine tendons with ethylene glycol diglycidyl ether has been reported for the repair of cruciate ligaments (Sung et al., 1996). The 1,4-butanediol diglycidyl ether (BD) has been reported as a crosslinking agent in the preparation of bioprosthetic valves (Zeeman et al., 2000). However, the fixation of porcine valves with BD caused immune response, foreign body reaction (proliferation of lymphocytes and macrophages) and calcification of implanted tissue using rats as animal model to levels similar to glutaraldehyde-fixed tissue, although low levels of cytotoxicity were reported (van Wachem et al., 2000). The combined treatment of BD and EDAC-dicarboxylic acid or detergents led to a reduction in calcification (implantation in rats) but not at significant levels (van Wachem et al., 1994). Therefore, it was concluded that the treatment with BD did not represents an alternative to glutaraldehyde to reduce the calcification of bioprosthetic valves (van Wachem et al., 1994). However, in another report the crosslinking of bovine perichardium and porcine aortic valves with triglycidylamine, a molecule of high polarity and solubility in water, resulted an improvement in biocompatibility (assessed using bovine aortic valve interstitial cells, human umbilical endothelial cells and rats artery smooth muscle cells) and resistance to calcification (subcutaneous implantation in rats) compared with glutaraldehyde-fixed tissues (Connolly et al., 2005). Furthermore, triglycidylamine-fixed tissues showed stable mechanical properties (Sacks et al., 2007) and optimal reduction of calcification when treatments included mercapto-aminobisphosphonate (Rapoport et al., 2007). It was hypothesized that the difference between these two results, which explored the chemistry of epoxy in the crosslinking of tissue, may be due to differences in water solubility, chemical heterogeneity and contamination with used epoxy residual reactants (Connolly et al., 2005).

4.4 Tissue crosslinking with diisocyanate

Bifunctional molecules capable of crosslinked proteins by urea bond formation after reaction between terminal isocyanate groups and ε-amino group of lysine residue have been explored. Such is the case of crosslinking of extracellular matrix proteins (elastin and fibronectin) with hexamethylene diisocyanate in dimethyl sulfoxide (Tirrell and Nowatzki, 2004). Similarly, the crosslinking of ovine skin collagen with hexamethylene diisocyanate has been reported. This crosslinking procedure was carried out in an aqueous medium including surfactants to increase solubility and promote the penetration of diisocyanate into the tissue (Olde Damink et al., 1995). Futhermore, the effects of the tissue crosslinking with hexamethylene diisocyanate and the effects of mixtures of water/isopropanol (50/50 and 0/100) as solvent on the thermal and biomechanical properties of bovine perichardium have been reported (Naimark et al., 1995). On the other hand, the stabilization of porcine perichardium has been achieved by the interaction of polyurethane oligomers containing isocyanate end groups (Loke et al., 1996). The interaction in organic media between perichardial tissue and polyurethane oligomers resulted in the increase of the denaturation temperature, a reduction in the content of lysine and a poor diffusion of polyurethane oligomers into the tissue (H&E staining). The crosslinking of bovine perichardium with polyurethane oligomers, EDAC and diphenyl phosphoric azide showed less cytotoxicity (assessed by a direct cytotoxicity test or Homsy test) than the tissue crosslinking with glutaraldehyde (Jorge-Herrero et al., 2005).

After these results, it is clear that diisocyanates are an alternative to glutaraldehyde in the preparation of bioprostheses. However, protein fixation with isocyanates has the disadvantage of using organic solvents. In addition, during the fixation in aqueous media, the crosslinking degree can be reduced due to competition of hydrolysis reactions. Therefore, the blocking reaction of isocyanate with bisulphite salts is an alternative in the preparation of water soluble isocyanates (Petersen, 1949). The protein crosslinking process with blocking isocyanates has the advantages of the use of aqueous media and reduced isocyanate toxicity (Mata-Mata et al., 2008). In this regard, the treatment of perichardial tissue with the carbamoylsulphonate blocked polyurethane prepolymers resulted in an increase of the in vitro tissue biostability (Mendoza-Novelo, 2011). The coating of collagen fiber network of perichardial tissue with polyurethane is shown in figure 8.

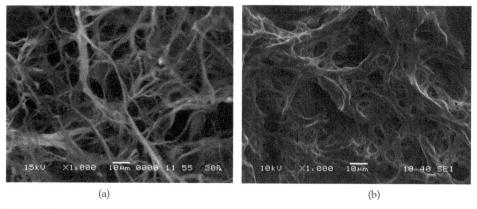

(a)   (b)

Fig. 8. SEM micrographs for bovine perichardium (a) native and treated with polyurethane prepolymers

4.5 Tissue crosslinking with naturally-derived compounds

Crosslinking agents of natural origin have also been explored in the tissue crosslinking. Such is the case of genipin, an aglycone or an iridoid glycoside, which can be obtained by enzymatic hydrolysis of the glucoside previously isolated from gardenia fruit. The stabilization of porcine perichardium (Sung et al., 1999) and acellular bovine perichardium (Sung et al., 2000) with genipin probably was achieved through cyclic structures. The crosslinking density for genipin-fixed tissue was similar to glutaraldehyde and ethylene glycol diglycidyl ether -crosslinked tissues. Moreover, the genipin-crosslinked porcine perichardium was less cytotoxic (fibroblasts) than glutaraldehyde-crosslinked tissue, whereas exhibiting the same tensile strength and resistance to enzymatic degradation (Chang et al., 2002). Furthermore, acellular bovine perichardium fixed with genipin showed capacity of angiogenesis (microvessel infiltration) after implantation in rats (Liang et al., 2004). Moreover, cell extraction with solutions of Triton™X-100 and the crosslinking with different concentrations of genipin were used to establish a relationship between the crosslinking degree and the degradation rate or the model of acellular tissue regeneration (Chang et al., 2004).

Polyphenolic compounds have also been investigated as natural agents of tissue stabilization, such as the proanthocyanidins from the family known as condensed tannins, which are essentially oligomers of flavonoids available in several fruits and vegetables. The stabilization of collagen with proanthocyanidins may involve the formation of hydrogen bond type interactions between the phenolic hydroxyl and amide carbonyls of the polypeptide chains.

The proanthocyanidin has a high affinity for proline-rich proteins, because this amino acid is a good hydrogen bond acceptor (Zhai et al., 2006). The proanthocyanidins can be used to crosslink collagen sponges with similar density and efficiency to glutaraldehyde but with reduced calcification after 6 weeks implantation in rats and it was reported to be 120 times less toxic to fibroblasts direct contact (Han et al., 2003). The proanthocyanidin crosslinking procedure was repeated in decellularized porcine aortic valves resulting in low toxicity to bovine aortic valve interstitial cells and in the stimulation of cell proliferation to low concentrations of this stabilization agent in the culture media (Zhai et al., 2009).

The stabilization of elastin in porcine aortas has been achieved by treatment with polyphenolic tannins, which is composed of a central molecule of glucose (hydrophobic core) and one or more galoil residues (hydrophilic shell) (Isenburg et al., 2006). Polyphenolic compounds were acetylated tannic acid, pentagaloil glucose, gallic acid and glucose. In this study, pentagaloil glucose treatment was the least toxic to fibroblasts (Isenburg et al., 2004). Also, the study revealed that polyphenolic hydroxyl groups are essential for the interaction between the tannic acid and elastin. The combination of tannic acid and glutaraldehyde rendered a biostable tissue with high resistance toward elastase and collagenase and low tendency to calcify (Isenburg et al., 2006).

The reuterin (β-hydroxypropionic acid) produced by *Lactobacillus reuteri* has been used in the fixation of porcine perichardium (Sung et al., 2002). The reuterin is soluble in water, with antimicrobial and antifungal activity. The properties of reuterin-fixed tissue are comparable to glutaraldehyde-fixed tissue in terms of amino group content, denaturation temperature, tensile strength and collagenase digestion resistance (Sung et al, 2003).

Microbial (mTG; *Streptoverticillium mobaraense*) and tissue (TG2; tTG) transglutaminases (protein-glutamine γ-glutamyltransferase, EC 2.3.2.13) have been explored in the crosslinking of collagen type I due to their ability to crosslink proteins through the ε-amino group of lysine and γ-carboxamide group of glutamine residue (Chen et al., 2005; Chau et al., 2005). The results indicated the efficiency of this crosslinking agents in terms of denaturation temperature, mechanical strength, low toxicity to fibroblasts (Chen et al., 2005) and an increase in osteoblasts and fibroblasts adhesion and proliferation compared to native collagen (Chau et al., 2005).

4.6 Other methods for the tissue stabilization

Others non-aldehydic crosslinking procedures have been proposed with the aims of prevent or mitigate tissue calcification. The disuccinimidyl glutarate (DSG) is another non- aldehyde alternative to tissue crosslinking. The process is carried out by the reaction between primary amino groups of tissue and NHS ester groups of DSG forming amide bonds with a length of five-carbon crosslinking and releasing NHS. The DSG crosslinked tissue was resistant to enzymatic degradation, exhibited low tendency to calcify and high temperature of denaturation. However, it was necessary to use dimethyl sulfoxide due to the insolubility of crosslinking agent in water (Pathak et al., 2000). In response to this drawback, a water soluble crosslinking agent has been used, i. e., the disuccinimidyl suberate. The presence of sulfonyl groups at the ends of the molecule conferred water solubility while retaining reactivity with amino groups by crosslinking chemistry similar to DSG, but with a length of 8 carbon intermediates. The tissue crosslinked under these conditions showed very low levels of calcium (0.2 mg/g of tissue) after 90 days of implantation in rats (Pathak et al., 2001). The crosslinking of collagen type I proposed for cartilage regeneration has also been achieved by the diimidoesters – dimethyl suberimidate (DMS). In this procedure, collagen amino groups react with DMS imidoester groups to form amidine groups and a length crosslinking of 8 carbons (Charulatha & Rajaram, 2003).

The stabilization of bioprosthetic tissue by filling the tissue interstitial spaces with polyacrylamide hydrogel resulted in the mitigation of tissue calcification in a rat study (Oosthuysen et al., 2006). Physical methods such as photo-oxidation (Khorn et al., 1997) or the use of ultraviolet radiation (Pfau et al., 2000) have also been proposed for the crosslinking of collagen-rich biomaterials. However, despite the increase in tissue shrinkage temperature, in some case the treated tissue did not show resistance to the proteins extraction (Moore et al., 1996).

4.7 Masking reactions

At this point it is important to distinguish between the effective formation of crosslinking sites, i. e., two reactive sites in collagen linked by a same molecule of crosslinking agent, and the masking of crosslinking, i.e., the reaction between a single end of bifunctional crosslinking agent and one reactive site of collagen. Table 6 shows the possible reactions of crosslinking and masking between collagen and difunctional crosslinking agents.

4.8 Glycosaminoglycans stabilization

Glycosaminoglycans present in both aortic valves and perichardium have been fixed to prevent the loss of these polysaccharides during the fixation of bioprosthetic valves. The sodium metaperiodate has been used for the stabilization of glycosaminoglycans in porcine aortic valves with the subsequent glutaraldehyde crosslinking (Vyavahare & Lovekamp, 2001). The stabilized porcine aortic valve showed compatibility and reduced calcification

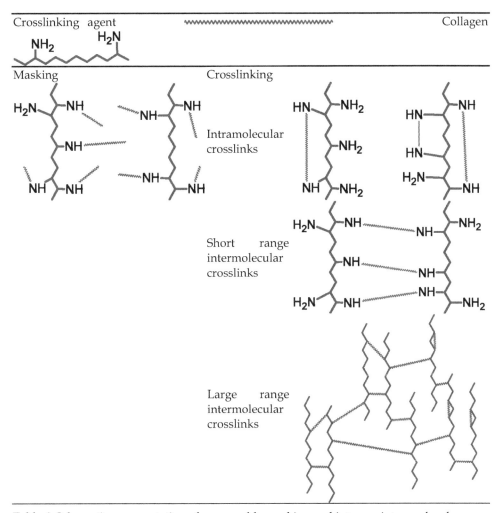

Table 6. Schematic representation of presumable masking and intra- or inter-molecular crosslinking structures

rates. Also it was reported the effectiveness of EDAC and neomycin (an inhibitor of the enzyme hyaluronidase) for the prevention of glycosaminoglycans loss (Ragharan et al., 2007; Shah & Vyavahare, 2008). The addition of exogenous glycosaminoglycans and the stabilization of endogenous glycosaminoglycans in ostrich perichardium reduced tissue calcification after implantation in rats, but slightly increased the presence of matrix-metalloproteinase at the implantation site (Arenaz et al., 2004).

5. Conclusions

Natural tissues from various sources can be used as biomaterials in the cardiovascular field after decellularization and fixation with various crosslinking agents. However, the current approaches based on surfactants and difunctional crosslinking agents can affect the structure, GAG content, biocompatibility and calcification potential of these tissues. Because of this, different postreatment methods have been suggested showing some improvements but until today, the vascular graft or heart valve obtained do not fulfill all the requirements for a long term use. In addition to this safety concerns, these treatments are not yet cost effective. Therefore, methods that preserve simultaneously the mechanical properties of collagen and the properties of the GAG matrix are desirable. Finally, those approaches that made use of enzymatic methods and low toxicity chemical such natural products and aminoacids as crosslinking agents seem promising alternatives.

6. References

Arenaz, B.; Martín-Maestro, M.; Fernández, P.; Turnay, J.; Olmo, N.; Senén, J.; Gil-Mur, J.; Lizarbe, M.A. & Jorge-Herrero, E. (2004). Effects of periodate and chondroitin 4-sulfate on proteoglycan stabilization of ostrich perichardium. Inhibition of calcification in subcutaneous implants in rats. *Biomaterials*, Vol. 25, pp. 3359-3368

Argese, E.; Marcomini, A.; Bettiol, C.; Perin, G. & Miana, P. (1994). Submitochondrial particle response to linear alkylbenzene sulfonates, nonylphenol polyethoxylates and their biodegradation derivates. *Environmental Toxicology & Chemistry*, Vol. 13, pp. 737-742

Carpentier, S.M., Shen, M., Chen, L., Cunanan, C.M., Martinet, B. & Carpentier, A. (2001). Biochemical properties of heat-treated valvular bioprostheses. *The Annals of Thoracic Surgery*, Vol. 71, pp. S410-S412

Carraway, K.L. & Koshland, D.E. (1972). Carbodiimide modification of proteins. *Methods in Enzimology*, Vol. 25, pp. 616-623

Cauich-Rodríguez, J.V. (2008). Materiales sustitutivos del tejido blando, In: *Biomateriales*, Sastre, R., de Aza, S. & San Román, J., (Ed.), pp. 359- 368, Faenza Editrice Iberica, Faenza RA, Italia

Chang, Y.; Liang, H.C.; Wei, H.J.; Chu, C.P. & Sung, H.W. (2004). Tissue regeneration patterns in a cellular bovine implanted in a canine model as a vascular path. Journal of Biomedical Materials Research, Vol. 69A, pp. 323-333

Chang, Y.; Tsai, C.C.; Liang, H.C. & Sung, H. W. (2002). In vivo evaluation of cellular and acellular bovine pericardia fixed with a naturally occurring crosslinking agent (genipin). *Biomaterials*, Vol. 23, pp. 2447-2457

Charulatha, V. & Rajaram, A. (2003). Influence of different crosslinking treatments on the physical properties of collagen membranes. *Biomaterials*, Vol. 24, pp. 759-767

Chau, D.Y.S.; Collighan, R.J.; Verderio, E.A.M.; Addy, V.L. & Griffin, M. (2005). The cellular response to transglutaminase-cross-linked collagen. *Biomaterials*, Vol. 26, pp. 6518-6529

Chen, R.N.; Ho, H.O. & Sheu, M.T. (2005). Characterization of collagen matrices crosslinked using microbial transglutaminase. *Biomaterials*, Vol. 26, pp. 4229-4235

Chen, W.; Kim, J.D.; Schoen, F.J. & Levy, R.L. (1994). Effect of 2-amino oleic acid exposure conditions on the inhibition of calcification of glutaraldehyde crosslinked porcine aortic valve. *Journal of Biomedical Materials Research*, Vol. 28, pp. 1485-1495

Connolly, J.M.; Alferiev, I.; Clark-Gruel, J.N.; Eidelman, N.; Sacks, M.; Palmatory, E.; Kronsteiner, A.; DeFelice, S.; Xu, J.; Ohri, R.; Narula, N.; Vyavahare, N. & Levy, R.J. (2005). Triglycidylamine Crosslinking of Porcine Aortic Valve Cusps or Bovine

Perichardium Results in Improved Biocompatibility, Biomechanics, and Calcification Resistance: Chemical and Biological Mechanisms. *American Journal of Pathology*, Vol. 166, pp. 1-13.

Courtman, D.W.; Pereira, C.A.; Kashef, V.; McComb, D.; Lee, J.M. & Wilson, G.J. (1994). Development of a perichardial acellular matrix biomaterial: Biochemical and mechanical effects of cell extraction. *Journal of Biomedical Materials Research*, Vol. 28, pp. 655-666

Dong, X., Wei, X., Yi, W., Gu, C., Kang, X., Liu, Y., Li, Q. & Yi, D. (2009). RGD-modified acellular bovine perichardium as a bioprosthetic scaffold for tissue engineering. *Journal of Materials Science: Materials in Medicine*, Vol. 20, pp. 2327-2336

Duan, X. & Lewis, R.S. (2002). Improved haemocompatibility of cysteine-modified polymers via endogenous nitric oxide. *Biomaterials*, Vol. 23, pp. 1197-1203

Duncan, A.C. & Boughner, D. (1998). Effect of dynamic glutaraldehyde fixation on the viscoelastic properties of bovine perichardial tissue. *Biomaterials*, Vol. 19, pp. 777-783

Everaerts, F.; Torrianni, M.; van Luyn, M.; van Wachem, P.; Feijen, J. & Hendricks, M. (2004). Reduced calcification of bioprostheses cross-linked via an improved carbodiimida based method. *Biomaterials*, Vol. 24, pp. 5523-5530

Flanagan, T.C. & Pandit, A. (2003). Living artificial heart valve alternatives: a review. European Cells & Materials, Vol. 6, pp. 28–45

Frost, M.C.; Reynolds, M.M. & Meyerhoff, M.E. (2005). Polymers incorporating nitric oxide releasing/generating substances for improved biocompatibility of blood-contacting medical devices. *Biomaterials*, Vol. 26, pp. 1685-1693

Gappa-Fahlenkamp, H. & Lewis, R.S. (2005). Improved haemocompatibility of poly(ethylene terephthalate) modified with various thiol-containing groups. Biomaterials, Vol. 26, pp. 3479-3485

Gappa-Fahlenkamp, H.; Duan X. & Lewis, R.S. (2004). Analysis of immobilized L-cysteine on polymers. *Journal of Biomedizal Materials Research*, Vol. 71A, pp. 519-527

García-Páez, J.M.; Jorge-Herrero, E.; Carrera-San Martín, A.; García-Sestafe, J.V.; Téllez, G.; Millán, I.; Salvador, J.; Cordón, A. & Castillo-Olivares, J.L. (2000). The influence of chemical treatment and suture on the elastic behavior of calf perichardium utilized in the construction of cardiac bioprostheses. *Journal of Materials Science: Materials in Medicine*, Vol. 11, pp. 273-2777

Gilbert, T.W.; Sellaro, T.L. & Badylak, S.F. (2006). Decellularization of tissues and organs. *Biomaterials*, Vol. 27, pp. 3675–3683

Giossis, G.; Yoshioka, S.A.; Braile, D.M. & Ramirez, V.D.A. (1998). The chemical protecting group concept applied in crosslinking of natural tissues with glutaraldehyde acetals. *Artificial Organs*, Vol. 22, pp. 210-214

Isenburg, T.C.; Karamchandani, N.V.; Simionescu, D.T. & Vyavahare, N.R. (2006). Structural requirements for stabilization of vascular elastin by polyphenolic tannins. *Biomaterials*, Vol. 27, pp. 3645-3651

Isenburg, T.C.; Simionescu, D. T. & Vyavahare, N. R. (2004). Elastin stabilization in cardiovascular implants: improved resistance to enzymatic degradation by treatment with tannic acid. *Biomaterials*, Vol. 25, pp. 3293-3302

Isenburg, T.C.; Simionescu, D.T. & Vyavahare, N.R. (2006). Tannic acid treatment enhances biostability and reduces calcification of glutaraldehyde fixed aortic wall. *Biomaterials*, Vol. 27, 1237-1245

Jee, K.S.; Kim, Y.S.; Park, K.D. & Kim, Y.H. (2003). A novel chemical modification of bioprosthetic tissues using L-arginine. *Biomaterials*, Vol. 24, pp. 3409-3416

Jobling, S. & Sumpter, J.P. (1993). Detergent components in sewage effluent are weakly estrogenic to fish – An in vitro study using rainbow-trout. *Aquatic Toxicology*, Vol. 27, pp. 361-372

Jorge-Herrero, E.; Fernández, P.; de la Tone, N.; Escudero, C.; García-Páez, J.M.; Buján, J. & Castillo-Olivares, J.L. (1994). Inhibition of the calcification of porcine valve tissue by selective lipid removal. *Biomaterials*, Vol. 15, pp. 815–820

Jorge-Herrero, E.; Fernández, P.; Escudero, C.; García-Páez, J.M. & Castillo-Olivares, J.L. (1996). Calcification of perichardial tissue pretreated with different amino acids. *Biomaterials*, Vol. 17, pp. 571-575

Jorge-Herrero, E.; Fonseca, C.; Barge, A.P.; Turnay, J.; Olmo, N.; Fernández, P.; Lizarbe, M.A. & García-Páez, J.M. (2010). Biocompatibility and calcification of bovine perichardium employed for the construction of cardiac bioprostheses treated with different chemical crosslink methods. *Artificial Organs*, Vol. 34, No. 5, pp. E168–E176

Jorge-Herrero, E.; Olmo, N.; Turnay, J.; Martín-Maestro, M.; García-Páez, J.M. & Lizarbe, M.A. (2005). Biocompatibility of different chemical crosslinking in bovine perichardium employed for the construction of cardiac bioprostheses. Proceedings of 19th European Conference on Biomaterials, pp. P4, Sorrento, Italy, Sept 11-15

Kidane, A.G.; Burriesci, G.; Cornejo, P.; Dooley, A.; Sarkar, S.; Bonhoeffer, P.; Edirisinghe, M. & Seifalian, A.M. (2009). Current developments and future prospects for heart valve replacement therapy. *Journal of Biomedical Materials Research Part B: Appl Biomater*, Vol. 88B, pp. 290-303

Khor, E. (1997). Methods for the treatment of collagenous tissues for bioprostheses. *Biomaterials*, Vol. 18, pp. 95-105

Langdon, S.E.; Chernecky, R.; Pereira, C.A.; Abdulla, D. & Lee, J.M. (1999). Biaxial mechanical/structural effects of equibiaxial strain during crosslinking of bovine perichardial xenograft materials. *Biomaterials*, Vol. 20, pp. 137-153

Lee, J.M. & Boughner, D.R. (1981). Tissue mechanics of canine perichardium in different test environment. *Circulation Research*, Vol. 49, pp. 533–544

Lee, J.M. & Boughner, D.R. (1985). Mechanical properties of human perichardium. Differences in viscoelastic response when compared with canine perichardium. *Circulation Research*, Vol. 55, pp. 475-481

Lee, J.M.; Boughner, D.R. & Courtman, D.W. (1984). The glutaraldehyde-stabilized porcine aortic valve xenograft. II. Effect of fixation with or without pressure on the tensile viscoelastic properties of the leaflet material. *Journal of Biomedical Materials Research*, Vol. 18, pp.79-98

Lee, J.M.; Edwards, H.H.L.; Pereira, C.A. & Samii, S.I. (1996). Crosslinking of tissue-derived biomaterials in 1-ethyl-3-(3-dimethylaminopropyl)-carbodiimide (EDC). *Journal of Materials Science: Materials in Medicine*, Vol. 7, pp. 531-541

Lee, J.M.; Haberer, S.A. & Boughner, D.R. (1989). The bovine perichardial xenograft: I. Effect of fixation in aldehydes without constraint on the tensile viscoelastic properties of bovine perichardium. *Journal of Biomedical Materials Research*, Vol. 23, pp. 457-475.

Lee, J.M.; Pereira, C.A. & Kan, L.W.K. (1994). Effect of molecular structure of poly(glycidy1 ether) reagents on crosslinking and mechanical properties of bovine perichardial xenograft materials. *Journal of Biomedical Materials Research*, Vol. 28, pp. 981-992

Lee, W.K.; Park, K.D.; Han D.K.; Suh, H.; Park, J.C. & Kim, Y.H. (2000). Heparinized bovine perichardium as a novel cardiovascular bioprosthesis. *Biomaterials*, Vol. 21, pp. 2323-2330

Li, S.T. (2007). Biologic biomaterials: tissue-derived biomaterials (Collagen). In: *Biomaterials*, Wong, J. Y. & Bronzino, J.D., (Ed.), pp. 7-1 – 7-22, CRC Press, ISBN-13: 978-0-8493-7888-1, 3Boca Raton, FL, USA.

Liang, H.C.; Chang, Y.; Hsu, C.K.; Lee, M.H. & Sung, H.W. (2004). Effects of crosslinking degree of an acellular biological tissue on its tissue regeneration pattern. *Biomaterials*, Vol. 25, pp. 3541-3552.

Liao, J.; Joyce, E.M. & Sacks, M.S. (2008). Effects of decellularization on the mechanical and structural properties of the porcine aortic valve leaflet. *Biomaterials*, Vol. 29, pp. 1065-1074

Lindop, G.B.M. (2007). Pathology of the Heart. *Surgery*, Vol. 25, pp. 187-197

Lovekamp, J. & Vyavahare, N. Periodate-mediated glycosaminoglycan stabilization in bioprosthetic heart valves. (2001). *Journal of Biomedical Materials Research*, Vol. 56, pp. 478–486

Loke, W.K.; Khor, E.; Wee, A.; Teoh, S.H. & Chian, K.S. (1996). Hybrid biomaterials based on the interaction of polyurethane oligomers with porcine perichardium. *Biomaterials*, Vol. 17, pp. 2163-2172

Mata-Mata, J.L.; Mendoza, D.; Alvarado, D.I. & Amézquita, F.J. (2008). Dispersiones acuosas de poliuretano bloqueadas, una alternativa como agente curtiente. *Polímeros: Ciência e Tecnología*, Vol. 18, pp. 138-143.

Matsagas, M.I.; Bali, C.; Arnaoutoglou, E.; Papakostas, J.C.; Nassis, C. & Papadopoulos, G. (2006). Carotid endarterectomy with bovine perichardium patch angioplasty: mid-term results. *Annals of Vascular Surgery*, Vol. 20, pp. 614–619

Mavrilas, D.; Sinouris, E.A.; Vynios, D.H. & Papageorgakopoulou, N. (2005). Dynamic mechanical characteristics of intact and structurally modified bovine perichardial tissues. *Journal of Biomechanics*, Vol. 38, pp. 761–768

Mendoza-Novelo, B. (2011). Caracterización fisicoquímica y biológica de un biomaterial cardiovascular basado en pericardio bovino y prepolímeros de poliuretano. Doctoral thesis. Universidad de Guanajuato, México, Marzo.

Mendoza-Novelo, B. & Cauich-Rodríguez, J.V. (2009). The effect of surfactants, crosslinking agents and L-cysteine on the stabilization and mechanical properties of bovine perichardium. *Journal of Applied Biomaterials & Biomechanical*, Vol. 7, No. 2, pp. 123-131

Mendoza-Novelo, B.; Ávila, E.E.; Cauich-Rodríguez, J.V.; Jorge-Herrero, E.; Rojo, F.J.; Guinea, G.V. & Mata-Mata, J.L. (2011). Decellularization of perichardial tissue and its impact on tensile viscoelasticity and glycosaminoglycans content. *Acta Biomaterialia*, Vol. 7, pp. 1241-1248

Monsan, P.; Puzo, G. & Mazarguil, H. (1975). Mechanism of glutaraldehyde protein bond formation. *Biochimie*, Vol. 57, pp. 1281-1292

Moore, M.A.; Chen, W.M.; Phillips, R.E.; Bohachevsky, I.K. & McIlroy, B.K. (1996). Shrinkage temperature versus protein extraction as a measure of stabilization of photooxidized tissue. *Journal of Biomedical Materials Research*, Vol. 32, pp. 209-214

Naimark, W.A.; Pereira, C.A.; Tsang, K. & Lee, J.M. (1995). HMDC crosslinking of bovine perichardial tissue: a potential role of the solvent environment in the design of bioprosthetic materials. *Journal of Materials Science Materials in Medicine*, Vol. 6, pp. 235-241

Nowatzki, P.J. & Tirrell, D.A. (2004). Physical properties of artificial extracellular matrix protein films prepared by isocyanate crosslinking. *Biomaterials*, Vol. 25, pp. 1261-1267

Ohri, R.; Hahn, S.K.; Hoffman, A.S.; Stayton, P.S &, Giachelli, C.M. (2004). Hyaruronic acid grafting mitigates calcification of glutaraldehyde-fixed bovine perichardium. *Journal of Biomedical Materials Research*, Vol. 70A, pp. 328-334

Olde Damink, L. H. H., Dijkstra, P. J., van Luyn, M. J. A., van Wachem, P. B., Nieuwenhuis, P. & Feijen, J. (1995). Crosslinking of dermal sheep collagen using hexamethylene diisocyanate. Journal of Materials Science Materials in Medicine, Vol. 6, pp. 429-434

Olde-Damink, L.H.H.; Dijkstra, P.J.; van Luyn, J.A.; van Wachem, P.B.; Nieuwenhuis, P. & Feijen J. (1995). Glutaraldehyde as a crosslinking agent for collagen-based biomaterials. *Journal of Material Science: Materials in Medicine*, Vol. 6, pp. 460-472

Oosthuysen, A.; Zilla, P.P.; Human, P.A.; Schmidt, C.A.P. & Bezuidenhout, D. (2006). Bioprosthetic tissue preservation by filling with a poly(acrylamide) hydrogel. *Biomaterials*, Vol. 27, pp. 2123-2130

Park, K.D.; Lee, W.K.; Yun, J.Y.; Han, D.K.; Kim, S.H.; Kim, Y.H.; Kim, H.M. & Kim, K.T. (1997). Novel anyicalcification treatment of biological tissues by grafting of sulphonated poly(ethylene oxide). *Biomaterials*, Vo. 18, pp. 47-51

Pathak, C.P.; Moore, M.A. & Phillips, R.E. (2002). Anticalcification treatments for fixed biomaterials. US patent, 6,479,079

Pathak, C. P.; Phillips, R. E.; Akella, R. & Moore, M. A. (2000). Evaluation of disuccinimidyl glutarate as a crosslinking agent for tissues. *Sixth World Biomaterials Congress Transactions*, pp. 288, Hawaii, USA

Pathak, C.P.; Phillips, R.E. & Moore, M.A. (2001). Water soluble tissue crosslinkers for bioprostheses. *Society for Biomaterials 27th Annual Meeting Transactions*, pp. 130, USA

Petersen, S. (1949). Niedermolekulare Umserzungsprodukte aliphatisher Diisocyanate. 5. Mitteilung über Polyurethane. *Annalen der Chemie*, Vol. 562, pp. 205-229.

Pfau, J.C.; McCall, B.R.; Card, G.L.; Cheung, D.T. & Duran, C.M.G. (2000). Lipid Analysis of Ethanol Extracts of Fresh Sheep Perichardial Tissue. *Sixth World Biomaterials Congress Transactions*, pp. 726, Hawaii, USA

Pfau, J.C.; Umbriaco, J.; Duran, C.M.G. & Cheung, D.T. (2000). Cross-linking of collagen in various states of hydration using UV irradiation. *Sixth World Biomaterials Congress Transactions*, pp. 718, Hawaii, USA

Raghavan, D.; Simionescu, D.T. & Vyavahare, N.R. (2007). Neomycin prevents enzyme-mediated glycosaminoglycan degradation in bioprosthetic heart valves. *Biomaterials*, Vol. 28, pp. 2861- 2868

Rapoport, H.S.; Connolly, J.M.; Fulmer, J.; Dai, N.; Murti, B.H.; Gorman, R.C.; Gorman, J.H.; Alferiev, I. & Levy, R. J. (2007). Mechanisms of the in vivo inhibition of calcification of bioprosthetic porcine aortic valve cusps and aortic wall with triglycidylamine/mercapto bisphosphonate. Biomaterials, Vol. 28, pp. 690-699

Sacks, M.S.; Hamamoto, H.; Connolly, J.M.; Gorman, R.C.; Gorman III, J.H. & Levy, R.J. (2007). In vivo biomechanical assessment of triglycidylamine crosslinked perichardium. Biomaterials, Vol. 28, pp. 5390-5398

Schmidt, C.E. & Baier, J.M. (2000). Acellular vascular tissues: natural biomaterials for tissue repair and tissue engineering. Biomaterials, Vol. 21, pp. 2215–2231

Schoen, F.J. & Levy, R.J. (1999). Tissue heart valves: current challenges and future research perspectives. Journal of Biomedical Materials Research, Vol. 47, pp. 439–75

Shah, S.R. & Vyavahare N.R. (2008). The effect of glycosaminoglycan stabilization on tissue buckling in bioprosthetic heart valves. Biomaterials, Vol. 29, pp. 1645-1653

Sheehan, J.C. & Hlavka, J.J. (1956). The use of water-soluble and basic carbodiimides in peptide synthesis. Journal of Organic Chemistry, Vol. 21, pp. 439-441

Sheehan, J.C. & Hlavka, J.J. (1957). The cross-linking of gelatin using a water-soluble carbodiimide. Journal of the American Chemical Society, Vol. 79, pp. 4528-4529

Soto, A.M.; Justicia, H.; Wray, J.M. & Sonnenschein, C. (1991). p-Nonylphenol, an estrogenic xenobiotic released from modified polystyrene. Environ Health Persp, Vol. 92, pp. 167-173.

Stacchino, C.; Bona, G.; Bonetti, F.; Rinaldi, S.; Della Ciana, L. & Grignani, A. (1998). Detoxification process for glutaraldehyde-treated bovine perichardium: biological, chemical and mechanical characterization. The Journal of Heart Valve Disease, Vol. 7, pp. 190-194

Sung, H.W.; Chang, Y.; Chang, W.H. & Chen, Y.C. (2000). Fixation of biological tissues with a naturally occurring crosslinking agent: fixation rate and effect of pH, temperature and initial fixative concentration. Journal of Biomedical Materials Research, Vol. 52A, pp. 77-87

Sung, H.W.; Chang, Y.; Chiu, C.T.; Chen, C.N. & Liang, H.C. (1999). Crosslinking characteristics and mechanical properties of a bovine perichardium fixed with a naturally occurring crosslinking agent. Journal of Biomedical Materials Research, Vol. 47A, pp. 116-126

Sung, H.W.; Chen, C.N.; Chang, Y. & Liang, H. F. (2002). Biocompatibility study of biological tissues fixed by a natural compound (reuterin) produced by Lactobacillus reuteri. Biomaterials, Vol. 23, pp. 3203-3214

Sung, H.W.; Chen C.N.; Liang, H.F. & Hong, M.H. (2003). A natural compound (reuterin) produced by Lactobacillus reuteri for biological-tissue fixation. Biomaterials, Vol. 24, pp. 1335-1347

Sung, H.W.; Shih, J.S. & Hsu, C. S. (1996). Crosslinking characteristics of porcine tendons: Effects of fixation with glutaraldehyde or epoxy. Journal of Biomedical Materials Research, Vol. 30, pp. 361-367

Tu, R.; Lu, C.L.; Thyagarajan, K.; Wang, E.; Nguyen, H.; Shen, S.; Hata, C. & Quijano, R.C. (1993). Kinetic study of collagen fixation with polyepoxy fixatives. Journal of Biomedical Materials Research, Vol. 23, pp. 3-9

van Wachem, T.P.B.; Brouwer, L.A.; Zeeman, R.; Dijkstra, P.J.; Feijen, J.; Hendriks, M.; Cahalan, P.T. & van Luyn, M.J.A. (2000). In vivo behavior of epoxy-crosslinked porcine heart valve cusps and walls. *Journal of Biomedical Materials Research Part B: Applied Biomaterials*, Vol. 53, pp. 18-27

van Wachem, T.P.B.; van Luyn, M.J.A; Olde Damink, L.H.H.; Dijkstra, P.J.; Feijen, J. & Nieuwenhuis, P. (1994). Biocompatibility and tissue regenerating capacity of crosslinked dermal sheep collagen. *Journal of Biomedical Materials Research*, Vol. 28, pp. 353-363

Vasudev, S.C. & Chandy T. (1999). Polyethylene glycol-grafted bovine perichardium: a novel hybrid tissue resistant to calcification. *Journal of Materials Science: Materials in Medicine*, Vol. 10, pp. 121-128

Vesely, I. (2005). Heart valve tissue engineering. *Circulation Research*, Vol. 97, pp.743–755

Vyavahare, N.; Hirsch, D.; Lerner, E.; Baskin, J.Z.; Schoen, F.J.; Bianco, R.; Kruth, Hs.; Zand, R. & Levy, R.J. (1997). Prevention of bioprosthetic heart valve calcification by ethanol preincubation. *Circulation*, Vol. 95, pp. 479-488

Yeung, B.K.S.; Chong, P.Y.C. & Petillo, P.A. (2001). Synthesis of glycosaminoglycans, In: *Glycochemistry*, Wang, P.C. & Bertozzi, C.R., (Ed.), pp. 425-492, Marcel Dekker, New York, U.S.A

Yoshioka, S.A. & Giossis, G. (2008). Thermal and spectrophotometric studies of new crosslinking method for collagen matrix with glutaraldehyde acetals. *Journal of Material Science: Materials in Medicine*, Vol. 19, pp. 1225-1223

Wei, H.J.; Liang, H.C.; Lee, M.H.; Huang, Y.C.; Chang, Y. & Sung, H.W. (2005). Construction of varying porous structures in acellular bovine pericardia as a tissue-engineering extracellular matrix. *Biomaterials*, Vol. 26, pp. 1905–1913

Whipple, E.B. & Ruta, M. (1974). Structure of aqueous glutaraldehyde. *Journal of Organic Chemistry*, Vol. 39, pp.1666-1668

Wiegner, A.W. & Bing, O.H.L. (1981). Mechanical and structural of canine perichardium. *Circulation Research*, Vol. 49, pp.807–814

Zeeman, R.; Dijkstra, P.J.; van Wachem, T.P.B.; van Luyn, M.J.A.; Hendriks, M.; Cahalan, P.T. & Feijen, J. (2000). The kinetics of 1,4-butanediol diglycidyl ether crosslinking of dermal sheep collagen. *Journal of Biomedical Materials Research*, Vol. 51, pp. 541-548

Zhai, W.; Chang, J.; Lin, K.; Wang, J.; Zhao, Q. & Sun, X. (2006). Crosslinking of decellularized porcine heart valve matrix by procyanidins. *Biomaterials*, Vol. 27, pp. 3684-3690

Zhai, W.; Chang, J.; Lü, X. & Wang, Z. (2009). Procyanidins-crosslinked heart valve matrix: Anticalcification effect. *Journal of Biomedical Materials Research Part B: Applied Biomaterials*, Vol. 90B, pp. 913-219

Zilla, P.; Brink, J.; Human, P. & Bezuidenhout D. (2008). Prosthetic heart valves: Catering for the few. *Biomaterials*, Vol. 29, pp. 385-406

Permissions

The contributors of this book come from diverse backgrounds, making this book a truly international effort. This book will bring forth new frontiers with its revolutionizing research information and detailed analysis of the nascent developments around the world.

We would like to thank Prof. Rosario Pignatello, for lending his expertise to make the book truly unique. He has played a crucial role in the development of this book. Without his invaluable contribution this book wouldn't have been possible. He has made vital efforts to compile up to date information on the varied aspects of this subject to make this book a valuable addition to the collection of many professionals and students.

This book was conceptualized with the vision of imparting up-to-date information and advanced data in this field. To ensure the same, a matchless editorial board was set up. Every individual on the board went through rigorous rounds of assessment to prove their worth. After which they invested a large part of their time researching and compiling the most relevant data for our readers. Conferences and sessions were held from time to time between the editorial board and the contributing authors to present the data in the most comprehensible form. The editorial team has worked tirelessly to provide valuable and valid information to help people across the globe.

Every chapter published in this book has been scrutinized by our experts. Their significance has been extensively debated. The topics covered herein carry significant findings which will fuel the growth of the discipline. They may even be implemented as practical applications or may be referred to as a beginning point for another development. Chapters in this book were first published by InTech; hereby published with permission under the Creative Commons Attribution License or equivalent.

The editorial board has been involved in producing this book since its inception. They have spent rigorous hours researching and exploring the diverse topics which have resulted in the successful publishing of this book. They have passed on their knowledge of decades through this book. To expedite this challenging task, the publisher supported the team at every step. A small team of assistant editors was also appointed to further simplify the editing procedure and attain best results for the readers.

Our editorial team has been hand-picked from every corner of the world. Their multi-ethnicity adds dynamic inputs to the discussions which result in innovative outcomes. These outcomes are then further discussed with the researchers and contributors who give their valuable feedback and opinion regarding the same. The feedback is then collaborated with the researches and they are edited in a comprehensive manner to aid the understanding of the subject.

Apart from the editorial board, the designing team has also invested a significant amount of their time in understanding the subject and creating the most relevant covers. They scrutinized every image to scout for the most suitable representation of the subject and create an appropriate cover for the book.

The publishing team has been involved in this book since its early stages. They were actively engaged in every process, be it collecting the data, connecting with the contributors or procuring relevant information. The team has been an ardent support to the editorial, designing and production team. Their endless efforts to recruit the best for this project, has resulted in the accomplishment of this book. They are a veteran in the field of academics and their pool of knowledge is as vast as their experience in printing. Their expertise and guidance has proved useful at every step. Their uncompromising quality standards have made this book an exceptional effort. Their encouragement from time to time has been an inspiration for everyone.

The publisher and the editorial board hope that this book will prove to be a valuable piece of knowledge for researchers, students, practitioners and scholars across the globe.

List of Contributors

Christiane E. Römer and Lothar Elling
Helmholtz-Institute for Biomedical Engineering, RWTH Aachen University, Germany

Leif Hermansson
Doxa AB, Sweden

Federica Chiellini and Andrea Morelli
Laboratory of Bioactive Polymeric Materials for Biomedical and Environmental Applications (BIOlab) UdR-INSTM – Department of Chemistry and Industrial Chemistry, University of Pisa, Italy

G. Fini, L.M. Moricca, A. Leonardi, S. Buonaccorsi and V. Pellacchia
La Sapienza/ Roma, Italy

Masaru Murata
Health Sciences University of Hokkaido, Japan

Toshiyuki Akazawa
Hokkaido Organization, Japan

Masaharu Mitsugi
Takamatsu Oral and Maxillofacial Surgery, Japan

In-Woong Um
Tooth Bank Co. Ltd, Korea

Kyung-Wook Kim
Dankook University, Korea

Young-Kyun Kim
Seoul National University Bundang Hospital, Korea

Jessem Landoulsi
Laboratoire de Réactivité de Surface, France
Université Pierre & Marie Curie -Paris VI, France

Michel J. Genet and Paul G. Rouxhet
Institute of Condensed Matter and Nanosciences – Bio & Soft Matter, Université Catholique de Louvain, Belgium

Karim El Kirat
Laboratoire de Biomécanique et Bioingénierie, France

Caroline Richard
Laboratoire Roberval, France

Sylviane Pulvin
Génie Enzymatique et Cellulaire, Université de Technologie de Compiègne, France

Gary D. Rayson and Patrick A. Williams
Department of Chemistry and Biochemistry New Mexico State University, Las Cruces, NM, USA

B.P. Zhang
National Engineering Laboratory for Carbon Fiber Technology, Institute of Coal Chemistry, Chinese Academy of Sciences, China
School of Materials Science and Engineering, Harbin Institute of Technology, China

Y. Wang and L. Geng
School of Materials Science and Engineering, Harbin Institute of Technology, China

Birzabith Mendoza-Novelo
Universidad Politécnica de Juventino Rosas, Mexico

Juan Valerio Cauich-Rodríguez
Centro de Investigación Científica de Yucatán, México